The Urban Book Series

Aims and Scope

The Urban Book Series is a resource for urban studies and geography research worldwide. It provides a unique and innovative resource for the latest developments in the field, nurturing a comprehensive and encompassing publication venue for urban studies, urban geography, planning and regional development.

The series publishes peer-reviewed volumes related to urbanization, sustainability, urban environments, sustainable urbanism, governance, globalization, urban and sustainable development, spatial and area studies, urban management, urban infrastructure, urban dynamics, green cities and urban landscapes. It also invites research which documents urbanization processes and urban dynamics on a national, regional and local level, welcoming case studies, as well as comparative and applied research.

The series will appeal to urbanists, geographers, planners, engineers, architects, policy makers, and to all of those interested in a wide-ranging overview of contemporary urban studies and innovations in the field. It accepts monographs, edited volumes and textbooks.

More information about this series at http://www.springer.com/series/14773

Fatemeh Farnaz Arefian
Seyed Hossein Iradj Moeini
Editors

Urban Change in Iran

Stories of Rooted Histories and Ever-accelerating Developments

Springer

Editors
Fatemeh Farnaz Arefian
Bartlett Development Planning Unit
University College London
London
UK

Seyed Hossein Iradj Moeini
Faculty of Architecture and Urbanism
Shahid Beheshti University
Tehran
Iran

ISSN 2365-757X ISSN 2365-7588 (electronic)
The Urban Book Series
ISBN 978-3-319-26113-3 ISBN 978-3-319-26115-7 (eBook)
DOI 10.1007/978-3-319-26115-7

Library of Congress Control Number: 2015955369

Springer Cham Heidelberg New York Dordrecht London

Printed on acid-free paper

Springer International Publishing AG Switzerland is part of Springer Science+Business Media
(www.springer.com)

Courtesy of Silk Cities Platform

Foreword

Gems on the Silk Roads

The story of this book is somehow unusual. It started in London in 2010, when Farnaz Arefian, an Iranian Architect and urban designer, at that time a doctoral researcher at the Bartlett Development Planning Unit (DPU), proposed to organise a small academic seminar around urban transformations in Iran. A few from us at the DPU got enthusiastic, as no recent international events or research programmes had discussed how Iranian cities expand, what key challenges they face, and how people and institutions are addressing them, in a country with close to 80 million inhabitants and that is renowned for its age-old multilayered urban cultures. As the preparation phase went on, we decided that the seminar should not only give an account of the transformations themselves, but examine critically these urban transformations in Iranian cities of all kind and size, and understand the rationale and underpinning logics explaining current situations, good or bad. To a large extent, this book brings new answers to these questions.

Soon after the conventional call for abstracts was launched, we realised that what was designed as a small high-profile international event was expanding swiftly, much beyond any original plans. Over 600 abstracts were received, way beyond the 40–50 expected. The scientific committee that gathered eminent specialists accepted generously to review ten times more proposals than originally planned, as their curiosity and interest were growing. The results were surprising: at least 150 abstracts deserved to be selected! A second observation from an academic point of view was that over 90 Iranian cities, of all sizes, and covering the whole Iranian territory were referenced, analysed, and in various cases critically examined. Two-thirds of the abstracts were coming from master's and doctoral students, essentially from Iranian universities, not only from Teheran but also from much smaller, less prestigious ones than those from the capital. To our opinion, this interest mirrors a fantastic desire to express and exchange ideas in an international scene, beyond the difficulties of writing in English. The seminar gave voice to a whole generation of young researchers, many of whom writing for the first time for an international academic event. In addition to this massive afflux from Iran, Iranian and non-Iranian scholars living outside Iran, sent quite a large number of fascinating proposals. They covered long civilisation and political periods, and not only those that followed the toppling of the Shah in 1979.

Unfortunately, the few papers selected here do not fully reflect the wealth of ideas, the bubbling of diverse opinions that were read. However, the book remains an excellent appetiser of different tastes and colours on the layering of different urban transformations that took place and that are taking place in the country.

As a way to 'face the flood,' give due visibility to a high number of promising papers, and stimulate the creativity of their authors, 150 of them were invited to deliver an exhibition panel that could be in Persian and English and that would conceptually and visually give an account of a longer paper. Again, this excellent complement that was exhibited during the London conference on November 8–9, 2012, is not included in the book, but deserves attention on the abstract proceeding of the conference, available at http://www.silkcities.org/data-bank.

The one strength that delineates the limits of the book comes from Iranian urban transformations seen through a kaleidoscopic view of Tehran, primarily of its spatial and social transformations and its waves of expansion. From chapter to chapter, the reader explores different realities that altogether sketch a unique view of the capital, as the laboratory of urban transformation in Iran. First, Tehran Bazaar is interpreted as a 'non-place,' following Marc Augé's anthropologic theories. Then, three iconic modern urbanism neighbourhoods are revisited and they allow understanding the metamorphosis of traditional Tehran into a modern metropolis, and at the same time, opening new paths of urbanisation in other urban centres. Detailed and refined analysis takes place at a much smaller scale, down to housing level, and explores the relationship between housing typology and urban morphology in order to understand the various transitions from courtyard houses to narrow row houses and to row apartments.

A review of urban images of Tehran through Iranian post-revolution movies enriches the vision and finds spaces through a perceptive question: How are urban spaces portrayed in movies and documentaries films staged and how do they link or not to other spaces in the city? This is an interesting question in relation to the fragmented growth of most metropolises that could find their unity only through representation and art, as a mean to reveal the city. Tehran is explored in another chapter as the scene of modernity that contributes to a long history of urban transformation of the capital.

Tehran 'call for spatial justice,' or the losing of natural resources and of agricultural land because of current models of urbanisation, or the illustration of old and new emerging spatial inequalities pattern, raises critical issues on the way the urban transformation has been taking place since the 1979 revolution. Such chapters bring unique points of view on urban Iran today and give to this book a unique position too. These critical visions are well complemented by an instigating Quranic reading of changes in today Islamic Iranian cities. Based on the philosophical pillars of Islam and its value systems, the work opposes the concept of modesty, reflected in traditional Islamic values and urban patterns, to the degree of immodesty of urban forms of the contemporary Islamic city, which de facto contradicts the philosophic pillars they are built upon.

These are only a few of the multiple visions of Tehran offered by the book. One should add the critical visit to Tehran satellite towns such as Hashtgerd New Town or some of the major regeneration projects. However, one of the interests of this collection lays on its capacity to drive the reader toward much less known and documented Iranian realities such as the Central Plateau's arid small cities such as Sabzevar or to Bam severely affected by the 2003 earthquake and to new towns, such as Shustar-e-No, one of many that would deserve a much greater critical review, for the challenges they raise and the serious problems that some of them are facing and that most probably will grow as time goes by.

One way to keep open the communication channels open beyond the international seminar was to carefully select a limited sample of papers. This book is a first account. A second mean was to launch a platform that would maintain and foster a dialogue on urban matters for the academic community within Iran, and just as importantly between Iran and its neighboring countries, and with far away scholars. The Silk Cities initiative, still modest, was an answer. Its name reminds a time when ancient Silk Road caravans were peacefully crossing Iranian ports and cities on their way from Istanbul to China. They brought wealth and new ideas, and were an engine for sharing knowledge. This book is in essence is a collection of gems found in some of the cities of the future silk roads of the twenty-first centuries, where knowledge on urban transformation will be one of the wealth to be shared and exchanged.

October 2014 Prof. Yves Cabannes

Acknowledgments

The preparation of this book owes much to the Silk Cities platform, an initiative for knowledge sharing and stimulating ideas for cities in the Middle East and Central Asia (http://www.silkcities.org; https://www.bartlett.ucl.ac.uk/dpu/silk-cities). It was also generously assisted by Maryam Eftekhar Dadkhah and Zohra Hasham, Asimina Paraskevepoulou at different stages of the book preparation, and Mitra Zelli for finding Hafiz's poetry. As previously mentioned, the contents of this book mostly build on revised and updated versions of a selection of papers from the international conference, Urban Change in Iran, hosted by the Bartlett DPU, UCL, on November 8–9, 2012. The conference created the momentum for the Silk Cities initiative. Academically chaired by Prof. Yves Cabannes, chair of Development Planning, the Bartlett Development Planning Unit, UCL, UK, it was principally organised by Fatemeh Farnaz Arefian, Ph.D., a researcher at the time at the Bartlett Development Planning, UCL, and director, Civitas Phoenix, UK, sponsored by UCL Institute for Risk and Disaster Reduction (IRDR), UK. The conference received significant support and input from independent academics and professionals as well as institutions from all around the globe, and received special attention from the UNESCO Director General, MS Irina Bokova, and was branded as 'a landmark event' by Oxford Brookes University. Institutional partnerships and supports are acknowledged on alphabetic order as the following: Centre of Excellence in Urban Design, Shahid Beheshti University, Iran; Civitas Phoenix Consulting, UK; Encyclopaedia Iranica, USA; Iranian Association of Urban Designers, Iran; Oxford Brookes University, UK; Technical University Berlin, Germany; The Society of Iranian Town Planners, Iran; This Big City (media partner), UK; UCLG Committee on Social Inclusion, Participatory Democracy and Human Rights; and Urban Design Group, UK.

Advisory supports and paper reviewers are acknowledged on alphabetic order as the following: Dr. Cyrus Alai, independent scholar, London, UK; Prof. David Alexander, University College London, UCL, UK; Dr. Camillo Boano, University College London, UCL, UK; Prof. Yves Cabannes, University College London, UK; Prof. Julio D. Davila, University College London; Mr. Farrokh Derakhshani, Aga

Khan Award for Architecture, Switzerland; Prof. Iraj Etessam, University of Tehran, Iran; Mr. Abdolhamid Eshragh, Encyclopaedia Iranica, France; Dr. Alireza Fallahi, Shahid Beheshti University (SBU), Iran; Prof. Nigel Harris, University College London, UCL, UK; Prof. Robert Hillenbrand, University of Edinburgh, UK; Dr. Bernard Hourcade, Centre National de la Recherche Scientifique (CNRS), France; Dr. Cassidy Johnson, University College London, UCL, UK; Dr. Hasan Karimian, University of Tehran, Iran; Prof. Ramin M. Keivani, Oxford Brookes University, UK; Prof. Hamid Majedi, Azad University, Iran; Prof. Ali Modarres, University of Washington Tacoma, USA; Dr. Farshad Nasrollahi, Technische Universität Berlin, Germany; Prof. Elke Pahl-Weber, Technische Universität Berlin, Germany; Dr. Parviz Piran, Shahid Beheshti University (SBU), Iran; Dr. Mina Saidi, Urban Observatory of Tehran and Cities of Iran, France; Prof. Ben Wisner, California State University, USA; Prof. Fulong Wu, University College London, UCL, UK.

Contents

Editors and Contributors

About the Editors

Fatemeh Farnaz Arefian is an academic consultant with extensive experience in private sector consultancy management, various-scale urban design and architectural projects, and research in Iran, the Middle East, and the UK. Her multidisciplinary Ph.D. research at the Bartlett DPU, UCL, was on development planning, focusing on organisational design and management for post-disaster reconstruction programmes. She was the initiator and principal organiser of the international conference at UCL in November 2012 (Urban Change in Iran), praised by the Director General of UNESCO. She is currently associated with the Bartlett DPU and the founding director of the Silk Cities Platform and director of Civitas Phoenix consultancy.

Seyed Hossein Iradj Moeini is a senior lecturer in architecture in Shahid Beheshti University (SBU), Tehran—where he has obtained his M.A.—and a practising architect in London. Having obtained his Ph.D. from the Bartlett School of Graduate Studies, UCL, on contemporary architectural theory and criticism, he is the author of numerous papers, has also taught in Yazd, Tehran Azad, and Shariati universities, and worked in a range of Iranian and British architectural practices in a professional capacity. A member of the Royal Institute of British Architects (RIBA) and the Iranian Comparative Arts Circle, he is also an architectural photographer and amateur musician.

Contributors

Fatemeh Farnaz Arefian The Bartlett Development Planning Unit, UCL, Silk Cities Platform, London, UK

Zahra Azizi Akaran Architects, University College London, London, UK

Naser Barakpou University of Art, Tehran, Iran

Manoochehr Dadashzadeh Science and Research Branch, Azad University, Tehran, Iran

Hassan Estaji University of Applied Arts Vienna, Vienna, Austria

Mahya Fatemi The Bartlett Development Planning Unit, University College London, London, UK

Hamed Goharipour Department of Urban and Regional Planning, Allameh Tabataba'i University, Tehran, Iran

Rana Habibi Department of Architecture, Urbanism and Planning, University of Leuven, Leuven, Belgium

Seyed Mohsen Habibi Faculty of Urbanism, Tehran University, Tehran, Iran

Farzaneh Haghighi Faculty of Architecture, Design and Planning, The University of Sydney, Sydney, Australia

Seavash Jamali Science and Research Branch, Azad University, Tehran, Iran

Maryam Kazemi Accademia delle Belle Arti di Brera, Milano, Italy

Ramin M. Keivani Oxford Institute for Sustainable Development, Department of Real Estate and Construction, Oxford Brookes University, Oxford, UK

Seyed Mahdi Khatami Faculty of Architecture, Design and Planning, The University of Sydney, Sydney, Australia

Azadeh Mashayekhi Faculty of Architecture, Spatial Planning and Strategies, Delft University of Technology, Delft, The Netherland

Bruno De Meulder Department of Architecture, Urbanism and Planning, University of Leuven, Leuven, Belgium

Parisa Mirsadeghi Shahid Beheshti University, Tehran, Iran

Seyed Hossein Iradj Moeini Faculty of Architecture and Urbanism, Shahid Beheshti University, Tehran, Iran

Hamidreza Rabiei-Dastjerdi Politecnico di Milano, Milano, Italy

Karin Raith University of Applied Arts Vienna, Vienna, Austria

Jenny Schmithals nexus Institute for Cooperation Management and Interdisciplinary Research, Berlin, Germany

Sabine Schröder nexus Institute for Cooperation Management and Interdisciplinary Research, Berlin, Germany

Homeira Shayesteh Bartlett School of Graduate Studies, UCL, London, UK

M. Reza Shirazi School of the Built Environment, Oxford Brookes University, Oxford, UK

Philip Steadman Bartlett School of Graduate Studies, UCL, London, UK

Michael Tawa Faculty of Architecture, Design and Planning, The University of Sydney, Sydney, Australia

Alireza Vaziri Zadeh Faculty of Engineering - Department of Architecture, KU Leuven, Leuven, Belgium

Chapter 1
Introduction

Fatemeh Farnaz Arefian and Seyed Hossein Iradj Moeini

> *Shiraz is a treasury of ruby lips, a quarry of beauty.*
> *Bankrupt jeweller that I am, it all makes me uneasy.*
> *So many drunken eyes I've seen, by God, in this town,*
> *I am so filled with cheer that I've abandoned wine.*
> *The town abounds with coy coquettes in each of the six*
> *Directions- I'm broke, else I'd buy all six*
>
> Hafiz, Persian Sufi and Poet, 14th century, translation from Lewisohn 2010?

Can we ever be able to respond to urban transformation while addressing sustainability, historical and traditional identity, socio-cultural concerns and social justice, economic growth and environmental challenges? Can design and planning or in general urban projects and programmes contribute to collective initiatives of urban development that seek to tap on the complexities of urban change in cities? These are our concerns.

This book examines and contributes in an understanding of the complexities of urban transformation in Iran, and thereby seeks ways of dealing with it. The overall approach is interpretive and qualitative, supported by quantitative research methods when needed. The book does not intend to provide a comprehensive statistical review or descriptive account of the subject. Instead, it provides qualitative critical reflections on contemporary practices of shaping cities and coping with urban change in Iran. It uses theoretical perspectives as tools to perceive, understand and explore how Iranian cities embrace urban transformation. The book mainly focuses on post-(1978) revolution Iran, going beyond the historicist account of traditional

F.F. Arefian (✉)
The Bartlett Development Planning Unit, UCL, Silk Cities Platform, London, UK
e-mail: ucfufaa@ucl.ac.uk; f.arefian@silkcities.org

S.H.I. Moeini
Faculty of Architecture and Urbanism, Shahid Beheshti University, Tehran, Iran
e-mail: h_moeini@sbu.ac.ir

F.F. Arefian and S.H.I. Moeini (eds.), *Urban Change in Iran*,
The Urban Book Series, DOI 10.1007/978-3-319-26115-7_1

urbanism and taking new perspectives such as exploring urban image through cinematic lenses, understanding urban phenomena through Islamic conceptual lenses and spatial justice, as well as organisational theory to the way urban programmes are formulated, organised and managed. Such insights shed light on both processes and products, connecting broad levels of policy making to very practical level of organising and implementing urban programmes. The process explores links between the physical, the social, and the political discussed from various perspectives, e.g. procedures of urban governance for achieving sustainability or the effects of planning processes on the country's urban identity. The analyses and debates offered in this book are deep enough for generalisations, suggesting models and informing similar studies of other cases, global or regional.

Rapid urbanisation and therefore intensive urban transformation are placing cities at the core of developmental agenda for the twenty first century. Urban transformation process presents both challenges and opportunities for thinking and acting out city futures (DPU n.d.). Regardless of who we are, an academic or a practitioner, a manager, a professional or another place-making actor, the current trend of urbanisation, especially in developing countries, presents us with overwhelming pressures and difficult decisions figuring out how to cope with urban change and flourish (Pieterse 2008). The persisting urban diversities render the 'one solution fitting all' as an elusive a dream as it ever was. Such diversities in developing countries stem from (and compound) broader structural developments such as cities' economic restructuring in the face of globalisation pressures, pressing environmental issues and power structures. In conjunction with local histories, cultures and social relations, these diversities produce highly differentiated, locally-specific patterns of urban development and change across the globe (DPU n.d.). The complexities of urban transformation, however, do not exist in isolation from its regional and global contexts. Seeking contextually sensitive approaches to deal with urban change, however, requires an exploration of the complexities of it within a given context in the first place.

Known as one of the oldest civilisations run by a government, many origins of urbanism can be traced back to Iran (or Persia, as it was called until 1935). Iranian architecture and urbanism has been a major influence in shaping an urban tradition now generically considered as that of the Islamic city: a tradition also resonating in some cities not considered as parts of the Islamic civilisation. According to UNDP (n.d.) '*Iran has been home to organized urban settlements since at least 4000 BC and even from those times the history of Iran has been intertwined with the history of the region as a whole*'. The country is home to some schools of urbanism such as the Isfahan School, which formed what is commonly known as Islamic urbanism. Such historic traditions can be traced in city centres and older parts of cities, in their urban morphology, architectural styles and the interrelations between natural and built environments.

Today Iran is a modern developing country. It is the Middle East-West Asia region's second biggest country after Saudi Arabia, and the second populated country, after Turkey, with an area of 1,648,000 km^2 and a population of almost

78 million (77,891,393) in November 2014 and an average population growth rate of 1.29 (Statistical Center of Iran 2014). Iran has adopted town planning regulations for over 50 years. Urban development programmes are guided and directed through comprehensive plans whose nature is inherited from the first comprehensive plan of Tehran in 1968. As Madanipour (2006) points out the 1968 comprehensive plan of Tehran (produced for the first time by a joint venture between an Iranian and an American consultancy) reflected the land-use based planning approach fashionable in the 1960s. Mainly based on American cities' patterns, it included a network of freeways and wide roads to connect disjointed areas, zoning as the basis for managing social and physical characteristics and introducing Floor Area Ratios for controlling development. Replicating such approach for other cities' comprehensive plans continued for 20 years with the municipalities being held responsible to implement these inflexible plans. The comprehensive plans treated traditional urban fabric in the same way as the newer parts. Plans determined the bulk and pocket volume of the buildings, including houses. The rate of urbanisation in Iran is comparable with Turkey and Egypt and less than countries in the south of the Persian Gulf, e.g. United Arab Emirates. This is, however, high in some respects: for example as a result of Iran's rapid urbanisation, 68.5 % of people lived in 1016 towns and cities in 2007. Tehran, Iran's capital is the most populated city in the region. The country has the greatest (7 out of 28) number of cities of more than one million in the region (The World Bank and United Cities and Local Governments 2008; United Cities and Local Governments 2010). Despite their noticeable geographic, climatic and economic diversity, Iranian cities follow the model of Tehran. The policies to respond to the urbanisation trend in Iran include those of developing new towns, inhabiting the excess population in existing cities, rehabilitating historic fabrics and creating public spaces, and flagship development projects. Following the discovery of oil in 1908, the country's economy shifted towards and oil-based economy. According to Human Development Report 2013, conducted by UNDP, Iran is one of the major oil exporting countries rated as a middle-income country; with an income per capita of $10,695 in 2013 (UNDP 2013).

Importantly, the country is the fourth natural-disaster prone country in the world, with many of its cities frequently subjected to severe damages and reconstruction throughout their histories. Traditionally and throughout the dry regions of central Asia and the Middle East, populations concentrate near dangerous earthquake-generating faults. This is in fact linked to their presence that is based on water as ancient settlements are usually located for reasons to do with water supply, access, strategic defence or controlling positions on trade routes, and these considerations are, in turn, often controlled by the land features created by earthquakes. To make matters worse, what were originally small villages have grown into towns, then cities, and now even mega-cities with several millions of people. But their growth has, in general, not been accompanied by any significant improvement in earthquake vulnerability. As a result, we should expect more disasters in this century, some of which far worse in mortality rates than those we have already seen (Jackson 2012). As an earthquake-prone country post-disaster reconstruction seems to be inevitable until after the country mainstreams measures and strategies for disaster risk reduction (DRR) towards safe construction and urban resilience. Until then, like many other countries in the region Iran must embrace for urban reconstruction.

The content of this book is mainly based on updated versions of a selection of papers originally peer reviewed and accepted for the international conference, *Urban Change in Iran*, hosted by the Bartlett Development Planning Unit (DPU), UCL, which focused on socio-cultural aspects of urban transformation in Iran along with the impacts of exposure to natural hazards on one hand, and the way in which they are dealt with on the other. Building on that, the book intends to bring together the knowledge of the dynamics of urban change and coping with that in the Iranian built environment context, in order to explore the ways in which the knowledge of the subject matter can inform the practice. The conference enjoyed contributions from many well-known scholars and practitioners in the field including Alireza Fallahi, Ali Modarres, Ben Wisner, Cassidy Johnson, David Alexander, Elke Pahl-Weber, Farrokh Derakhshani, Goran Car, Iraj Etessam, James Jackson, Jennifer Robinson, Klaus Rückert, Mohammad Chaichian, Nigel Harris, Parviz Piran, Ramin Keivani, Uwe Schäfer and Yves Cabannes. The conference, however, created a platform to give voice to the emerging generation of scholars who represent fresh insights into the subject matter: a voice echoed in this book.

Part I of the book deals with the question of the historic context for urban transformation in Iran. Chapters of this part discuss historic pre-modern Iranian cities, from water based urban formation to the bazaar as the traditional urban structure. It continues with discussing the emergence of modernity, manifested by flagship urban projects that in turn influenced the future of Tehran. It is followed by contemporary practices of city building that is further explored through cinematic lenses. Hassan Estaji and Karin Raith use Sabzevar as an example of water-based urban formation to demonstrate the correspondence between the urban form and the traditional irrigation system, the qanats. Farzaneh Haghighi revisits the Tehran Grand Bazaar using the concept of 'non-places' to unveil multiplicities of meanings in the bazaar. Rana Habibi, Bruno De Meulder and Mohsen Habibi review three mid-20th century residential districts in modern Tehran, to explore how Tehran has transformed into a modern city. Finally, Hamed Goharipoor explores associations between the city and cinema. The chapter sees Tehran as the main location for Iranian films and reviews how evolving views about Tehran are represented in post-revolution Iranian films.

Part II deals with the interconnections and relationships between the society and city formation. Seyed Mahdi Khatami uses Quranic concepts as a basis for his analytical framework particularly focusing on the concept of modesty in order to understand contemporary cities in the Islamic world for which Iran is an example. He links deeply-rooted cultural beliefs with practicalities of city life through urban projects and programmes. The relationship between city formation and urban transformation and people/society is bilateral. Although urban physical change can be perceived as a manifestation of deeper societal transformations, practical decisions on urban projects and programmes can in turn enhance or restrict socio-cultural aspects of urban life. Zahra Azizi and Mahya Fatemi examine crucial roles of urban design and planning towards promoting or thwarting just societies. They take a socially-oriented perspective to demonstrate the ongoing injustice in global cities and see Tehran as a case in the point. At a more practical level, Sabine Schröder and Jenny Schmithals present a fresh example of people's participation in design stage of new towns as an effective way of making sustainable places.

Part III engages the reader with a crucial question: how to achieve urban continuity while dealing with urban change? This is a concern for urban designers, planners, urban development practitioners and place makers, and of course urban managers, and a particularly grave issue for Iran: a country proud of her contributions to the emergence of what is now called Islamic cities. Given such background, Azadeh Mashayekhi reviews two urban projects for contemporary Tehran and explores the strong associations between socio-political transformation of Tehran and the built environment. While the reconciliation of traditional and contemporary city formation is a concern for many, Reza Shirazi revisits an urban project considered as avant-garde at a time exemplified by desires for new towns, the New Shushtar (Shushtar-e-No), and examines the gaps between the architect's social ideals and the real life and emerging political turbulences that ended it up in a dystopian situation. Parisa Mir-Sadeghi links urban continuity with the relation between the natural and the built environments, reviews the consequences of Tehran's rapid development from the greenery point of view, and shows how this destruction of natural reserves continues as the city sprawls out of its original boundaries. The environmental qualities, which promoted locating the city in the first place, is seen as at odds with environmental sustainability within the broader system of urban ecology.

Part IV deals with the question of the links between city formation and maintenance with its broader context and influential aspects. Naser Barakpou and Ramin Keivani discuss sustainability as a major discourse in urban studies, and examine the relationship between urban governance and sustainable urban development in Iran. Looking at deteriorating environmental conditions of Iranian cities, they argue that urban governance models need to replace current centralised governing models if any improvement is to be achieved. Hamidreza Rabiei-Dastjerdi and Maryam Kazemi compare spatial configurations during the contemporary history of Tehran. In doing so they first try to understand whether or not Tehran is a global city, and then employ planning indicators to argue that the traditional north-south divide is giving way to a new west-east divide under new social formations. Homeira Shayesteh and Philip Steadman discuss how architectural and urban design codes stemming from planning requirements for vehicular access and standardised day-lighting shaped new urban block formations. They use three examples of contemporary Tehran districts to explore a significant shift in housing typology in many Iranian cities in their move away from courtyards to row blocks and yards.

Part V deals with the question of how urban projects and programmes play roles in urban transformation at the field. A city can be seen as a portfolio of interconnected urban projects and programmes, while having different life cycles, and create a vibrant continuous flow of urban change. Siavash Jamali investigates the effects of Iran's first baby boom after 1978 and looks at pressures of the baby boomer generation in Iran on housing and how state solutions such as large-scale affordable housing projects 'Maskan-e-Mehr', failed to solve the problem as they just focus on construction and not community building. While cities as engines for economic growth must be able to generate income to flourish, the planning system in Iran limits urban managers' powers to do this (Harris 2012), and attempts for facilitating this through large-scale property-led redevelopment and regeneration programmes, as Alireza Vaziri Zadeh examines, have had consequences. Using two examples in Mashhad and Tehran, this chapter examines the property-driven approach and argues that the failure of mega-property projects is because of their property-driven approach, social polarisation and their state-led attitude with a lack of institutional capacity. Concluding this part, in a paper not originally presented in the conference, Fatemeh Farnaz Arefian discusses organising urban programmes as middle scale organisational structures which pursue certain objectives within the very challenging context of post-disaster urban reconstruction. She examines how the post-disaster housing reconstruction programme in Bam was organised and managed from the perspective of organisation theory.

The above examinations unfold complexities that can in turn indicate directions for further multidisciplinary studies in this field of city formation and maintenance. It is hoped to act as a first step in promoting contextually sensitive approaches in urban development in Iran.

References

Harris N (2012) Thoughts on Iranian cities in economic globalisation. Presented at international conference, urban change in Iran, UCL, London, 8–9 Nov 2012

Jackson J (2012) Geographical aspects of earthquake vulnerability in Iran. Presented at international conference, urban change in Iran, UCL, London, 8–9 Nov 2012

Lewisohn L (2010) 1-Socio-historical and literary contexts: Hafiz in Shiraz (trans: Hafiz, 14th century). In: Lewisohn L (ed) Hafiz and the religion of love in classical Persian poetry. I.B. Tauris

Madanipour A (2006) Urban planning and development in Tehran. Cities 23:433–438

Pieterse EA (2008) Urbanisation trends and implications. In: Pieterse EA (ed) City futures: confronting the crisis of urban development. Zed Books, London, pp 16–38

Statistical Center of Iran (2014) Iran, main indicators. http://www.amar.org.ir/Default.aspx?tabid=1241. Accessed 13 Nov 2014 (WWW Document)

The Bartlett Development Planning Unit (n.d.) Urban transformations. The Bartlett Development Planning Unit. http://www.bartlett.ucl.ac.uk/dpu/research/urban-transformation. Accessed 13 Oct 2014 (WWW Document)

The World Bank and United Cities and Local Governments (2008) Decentralisation and local democracy. First Global Report United Cities and Local Governments, Barcelona

UNDP (2013) Iran Extract from Human Development Report. Available from:http://www.undp.org/content/dam/iran/docs/Publications/Inclusive%20Growth%20&%20Development/IR%20of%20Iran%20Extract%20in%20HDR%202013%20ENG.pdf. Accessed 15 Feb 2014

UNDP (n.d.) About Iran. UNDP in Iran. http://www.ir.undp.org/content/iran/en/home/countryinfo/#Introduction. Accessed 13 Oct 2014 (WWW Document)

United Cities and Local Governments (2010) Local government and finance: the challenges of the 21st century. Second Global Report on Decentralisation and Local Democracy, Barcelona

Part I
Pre-modern and Contemporary Urbanism

Chapter 2
The Role of Qanat and Irrigation Networks in the Process of City Formation and Evolution in the Central Plateau of Iran, the Case of Sabzevar

Hassan Estaji and Karin Raith

Abstract The main cities of the central plateau of Iran are mostly located on the outer edge of the region. This distribution is determined by climatic impacts. Due to scarce precipitation and rapid evaporation of water for more than six months each year, the region lacks permanent rivers, and for that reason it was initially hard to establish permanent settlements. The population had to move between mountains and plains seasonally, until the invention of 'qanats' changed the way of life and settlement. By means of these underground aqueducts water was funnelled from mountainous areas and aquifers to lower lands. Alluvial fans could be opened up to settlement and an agrarian civilisation evolved. The qanat system is one of the important influences on the location and morphology of desert cities. By comparing old maps of the Sabzevar qanats and water routes with maps of historical streets and alleys in Sabzevar we found that the urban development pattern of Sabzevar is based on the transformation of farming lots to urban residential spaces. When in the course of urbanisation the fields were turned into houses with gardens, main streets were laid out along the course of main qanats and alleys along subsidiary channels. The resulting street pattern corresponds with the old irrigation network and, interestingly, the hierarchy of streets follows the hierarchy of the water distribution network.

Keywords Qanat · Irrigation network · Iranian cities · Urban morphology · Sabzevar

---

H. Estaji (✉) · K. Raith
University of Applied Arts Vienna, Vienna, Austria
e-mail: estaji@student.uni-ak.ac.at

K. Raith
e-mail: karin.raith@uni-ak.ac.at

F.F. Arefian and S.H.I. Moeini (eds.), *Urban Change in Iran*,
The Urban Book Series, DOI 10.1007/978-3-319-26115-7_2

2.1 Introduction

This chapter addresses the evolution and transformation processes of cities on the central plateau of Iran specially focusing on Sabzevar. We cannot investigate cities without considering their environments. 'The primary physical factors for the development of Central Iranian cities have been the hostile climate and the shortage of water within the bowl-shaped physiography of the Iranian Plateau' (Kheirabadi 2000: 1). This chapter examines the role of a specific type of water supply—the qanat system—in the formation and morphology of Iranian desert cities.

2.2 Iranian Desert Cities

Dasht-e Kavir and Dasht-e Lut, Iran's two largest deserts, which together amount to one seventh of the total area of the country, are located in the central plateau. The region is characterised by a typical desert climate with cold winters and hot and dry summers, low relative humidity (about 60 % in winters and 20 % in summers) and an average annual precipitation between 150 and 300 mm (Ghobadian 2009: 24). The main cities of the central plateau of Iran are mostly situated on the outer edge. This distribution of the cities within the territory is determined by climatic conditions. Due to scarce precipitation and rapid evaporation of water prevailing for more than six months of the year, the region lacks permanent rivers, and therefore it was hard to establish permanent settlements. A society of livestock breeders had to move between mountains and plains seasonally in order to feed themselves and their animals, but the invention of the 'qanat' (from a Semitic word meaning 'to dig', in Persian 'Kariz') changed living conditions. By means of these underground aqueducts water was funnelled from mountainous areas and aquifers to lower lands. Alluvial fans could be opened up to settlements, and agrarian civilisations evolved. The qanat became a crucial element of the habitat. The nomadic herding was largely replaced by the settled farming. Nadji (1973: 938–939) states that the sedentary civilisation in the central plateau of Iran was evolved by the qanat (Kariz) technology and calls it '*Kariz's Civilisation*'. In his opinion, over the past centuries the qanat system has played a key role in various aspects of civilisation in this region.

2.3 The Structure of Qanat

A qanat is a water supply system that taps groundwater reserves and channels them to settlements and agricultural areas located at a lower level. It consists of a slightly inclined underground gallery and a row of vertical shafts giving access to the low-lying tunnel. Usually 'the first shaft (mother well) is sunk[en] … into an alluvial fan to a level below the groundwater table' (International Centre on Qanats and Historic Hydraulic Structures (ICQHS 2012).

The length of a qanat depends on various factors: the topography, the location of the aquifer and the material of underground layers. It is between a few hundred metres up to several tens of kilometres. According to the definition of the ICQHS (2012) each qanat consists of three main sections: *water production* (the section where underground water resources are developed), *water transport* and *water use*. The most important parts of a qanat are described as follows (Fig. 2.1):

- *Gallery*: The gallery serves both water production and transport, and extends over both sections, i.e. from the groundwater recharge zone to the zone where the water is used. It is a tunnel which taps a water bearing stratum (aquifer) and channels the water to the surface over a long distance. The extremely low gradient [maximum 1:1000 or 1:1500 according to English (1968: 173)] prevents the gallery from being eroded by rapidly flowing water.
- *Shaft wells*: are the vertical funnels or ducts which are dug at intervals between 20 and 200 m (depending on the depth of the gallery) while the qanat is being constructed. They are used to remove excavation material and are essential for ventilating and maintaining the qanat.
- *Mother well*: The mother well is the shaft located closest to the aquifer at the beginning of the qanat. As such, it is also the deepest of all the shafts along the gallery. Mother wells can have a depth of up to 300 m (the qanat of Gonabad).

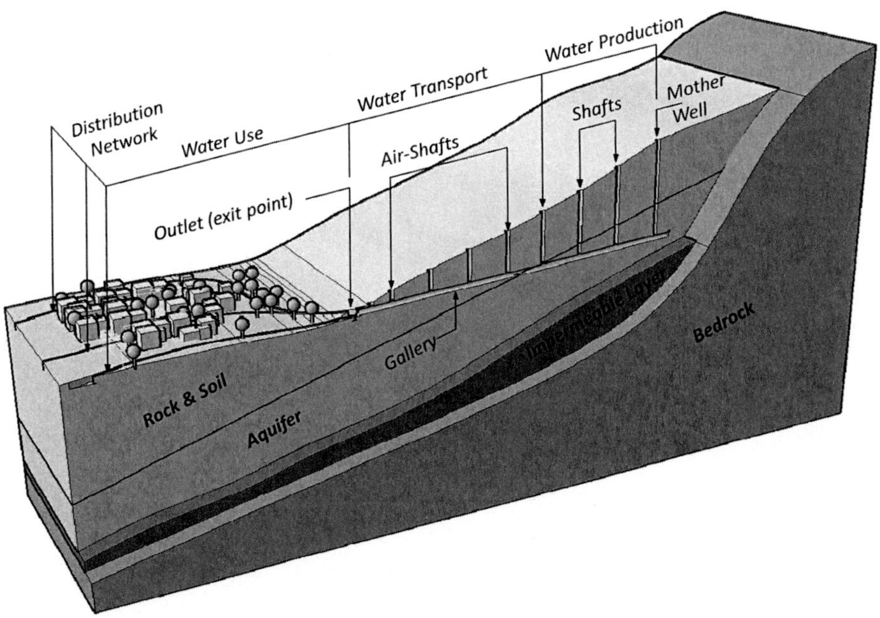

Fig. 2.1 3D cross-section of a typical qanat, illustration by authors, data from ICQHS (2012)

- *Exit point of the qanat*: At the exit point of the qanat or '*Mazhar*' (literally 'where the water appears') the gallery meets the surface.
- *Farm*: Agricultural areas located lower than the *Mazhar* are supplied with water via surface channels. 'The extent of the cultivated area depends on several factors such as the qanat discharge, soil quality, soil permeability, local climatic conditions, etc.' (ICQHS 2012).

2.4 Human Settlements and the Geographical Distribution of Qanats in Iran

'The qanat technology was known in Iran by the sixth century BC, when Indo-Iranians began to settle as agriculturists' (English 1998: 196). To correspond with positioning necessities of qanats almost all large towns and early settlements on the central plateau of Iran are situated in the plains between high mountains and the desert, on the outer edge of the central plateau of Iran. The statistics and the map of the geographical distribution of qanats in Iran clearly show this pattern (Fig. 2.2). By providing water for the fertile downstream lands qanats facilitated the development of Iranian cities in the plains of desert regions. ICQHS (2009: 5) states:

> In the years 1984–1985 the ministry of energy took census of 28038 qanats whose total discharge was 9 billion cubic metres. In the years 1992–1993 the census of 28054 qanats showed a total discharge of 10 billion cubic metres. 10 years later in 2002–2003 the number of the qanats was reported to be 33691 with a total discharge of 8 billion cubic metres.

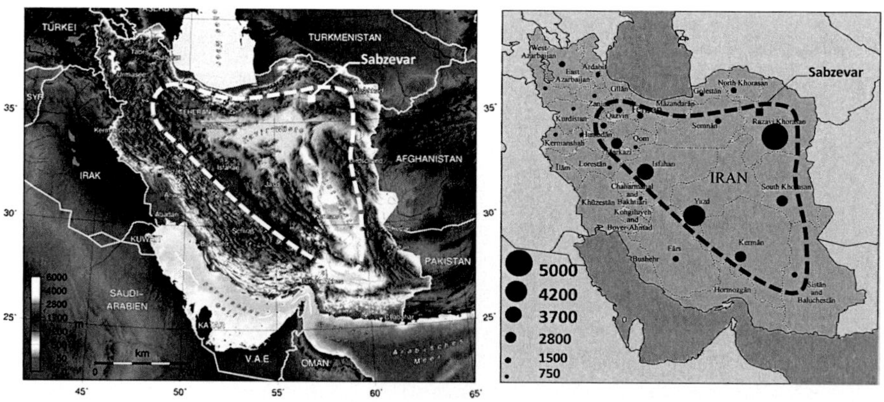

Fig. 2.2 *Left* Topographic map of Iranian central plateau (World of Maps 2012). *Right* The number of qanats in 2005 in each province, based on Iranian qanat database (Iran Hydrology 2005), blank map: (Sankakukei 2012)

According to the latest statistics (Semsar Yazdi 2010: 7), 'Iran is currently benefiting from 36,888 active qanats producing some 7 billion cubic metres of groundwater and forming around 11 % of the aquifer discharge which is annually mined across the country.'

2.5 Sabzevar

Sabzevar is located in the northeast of Iran, south of the Sabzevar Mountains (also known as Siah-Kuh or Joghatay Mountains) on the outer edge of the Central Plateau (Fig. 2.2). These mountains lie 80 km south of the East Alborz. The city and the neighbouring villages are situated between the mountain range and the Salt Desert, parallel to Sabzevar Mountains. According to the last statistics reported by the Ministry of Agriculture (Iran hydrology 2005), there are 864 qanats in Sabzevar and the surrounding villages. The position of the settlements and the main qanats confirms that Sabzevar and its surrounding rural settlements were established according to the qanat system. The distance between the mountains and the settlements is between 10 and 15 km depending on the length of the main qanats (Fig. 2.3).

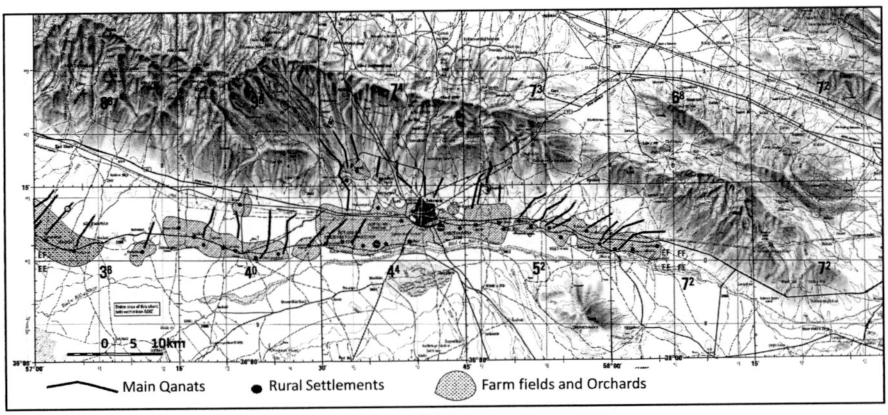

Fig. 2.3 Distribution of main qanats in the Sabzevar region, base map: (National Imagery and Mapping Agency 1999). (This product was developed using materials from the United States National Imagery and Mapping Agency and are reproduced with permission. This product has neither been endorsed nor authorised by the United States National Imagery and Mapping Agency or the United States Department of Defense)

2.6 The Qanat Network

> The built environments of most alluvial fan towns and villages on the Iranian Plateau are aligned along the major watercourses (shahjub) that run from the mouth of the qanat down slope through the length of the settlement (English 1998: 198).

The hierarchy of the irrigation network is as follows (Fig. 2.4):

1. Main irrigation channels: main water streams usually run in parallel with the main slope of the land.
2. Sub-main irrigation channels (link channels): flowing into fields following land topographies.
3. Subsidiary channels: branching from main networks. These smaller streams irrigate the gardens and farms and provide water for 'water storages' in the houses.

In the dry climate, water ownership is more important than the ownership of the land. The qanat network allows land owners to buy water from several qanats. This flexibility makes the irrigation network a sustainable system. If one of the qanats

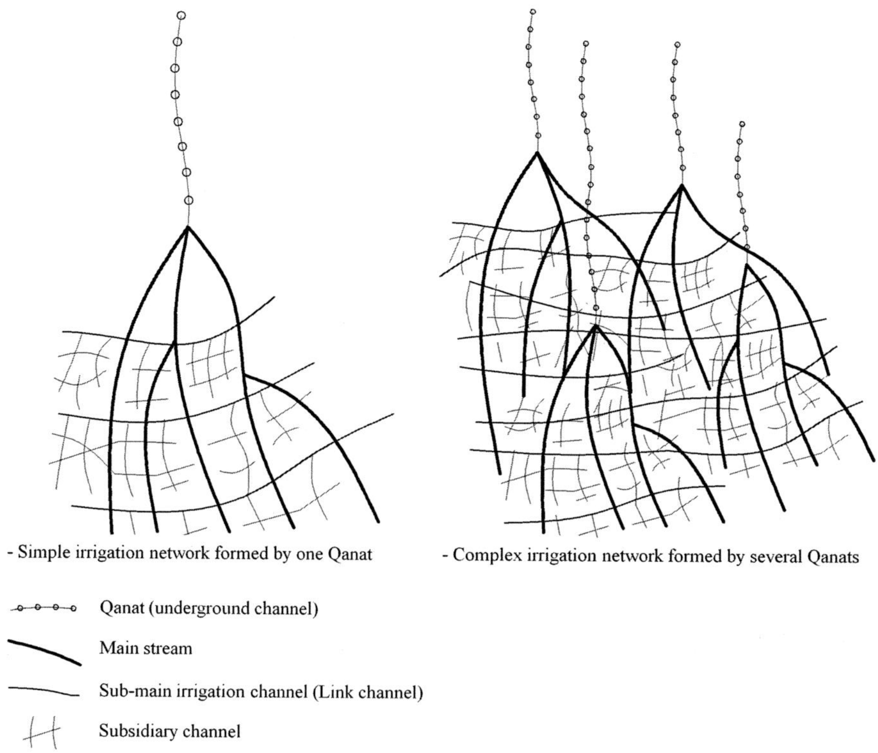

Fig. 2.4 Irrigation network (Authors 2012)

dries up or needs repair, this organic network enables farmers to carry water from another qanat into their farms and gardens via link channels (sub-main irrigation channels).

2.7 The Role of the Qanat in the City's Morphology

For purposes of repair and maintenance of the qanat and irrigation networks, and in order to make gardens and farms accessible, ancient Iranians made narrow paths for humans and animals on one or both sides of the channels. The width of these paths depends on water discharge rate and the hierarchy of the qanat network. The primary function of qanats and irrigation systems was to supply water for household consumption and agriculture but as a side-effect a spatial pattern was created which played a formative role in the city layout and urban evolution. The water supply system, therefore, is one of the important influences on the morphology of desert cities. As an Iranian desert city Sabzevar was also formed according to this pattern: a point mentioned by Kheirabadi (2000: 33) in his book. For a more detailed study, two maps of the old Sabzevar irrigation network and the historic streets and alleys were prepared (Fig. 2.5). By comparing these maps, it turns out that the urban pattern of the old Sabzevar matches the irrigation network. When in the course of urbanisation farming lots were built on and turned into houses with gardens, main streets were laid out along the course of main irrigation channels and alleys along subsidiary channels. The resulting street pattern corresponds with the irrigation network and. interestingly, the hierarchy of streets follows the hierarchy of the water distribution net. This map shows the final years (around 1950) of the qanat system still fully functioning in Sabzevar. At that time three main and around 15 small qanats were included in the system providing water for 30,000 people.

In the early twentieth century a transformation from an agrarian to urban society took place. A growing wealth together with the development of fossil energies led to a sharp increase in motor traffic. As a result of population growth, the demand for drinking water rose. Cities expanded into suburban areas, and overran gardens. Surface channels were replaced by underground water pipes for sanitary reasons. The vital vessels of running water on the surface lost their life and gardens dried up. These developments happened about 60 years ago. Nevertheless the 'dead network' and the dried gardens and farms kept their formative role for urban development. The network of the irrigation system which was already accompanied by public spaces changed into new streets and alleys and dried farms and gardens were divided into smaller lots for residential buildings (Fig. 2.5). The first law concerning the 'widening and development of streets' in Iran approved by the Iranian parliament in 1933, speed up these changes.

The municipalities made a master plan with the objective of facilitating traffic. In the first years of this urbanisation the traffic density was still relatively low, and therefore problems could be solved by slightly modifying the dried irrigation network (Fig. 2.6).

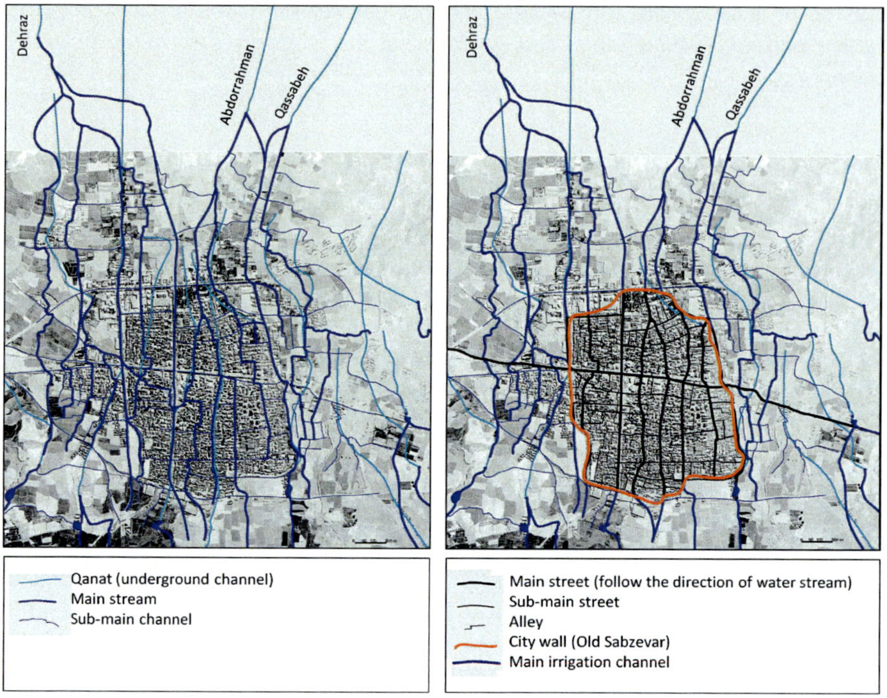

Fig. 2.5 Old Sabzevar irrigation network and the position of old Sabzevar and city wall based on the 1956 Sabzevar map. (Authors based on the map by Zanganeh (2003a), the base map reproduced by permission of Yaghoub Zanganeh)

This picture shows the transformation process of farming lots to urban residential spaces in Sabzevar (the southwest of the old city is shown as an example). The last part of the old city wall can be seen clearly in the 1956 aerial photo—it was demolished and replaced with a new street in 1964.

Until 1975 the irrigation network determined the street pattern even after losing its original function. After that the organic network could not cope with the traffic any longer due to the increasing numbers of cars. In the last decades the traditional order of the city was abandoned. The authorities and urban designers chose the simplest solution for the traffic problem: starting in 1975, they cut a rectangular grid of new wide streets into the urban fabric of Sabzevar—ignoring the centuries-old structure of the city. These cuts can be seen in the Sabzevar aerial photo (Fig. 2.6).

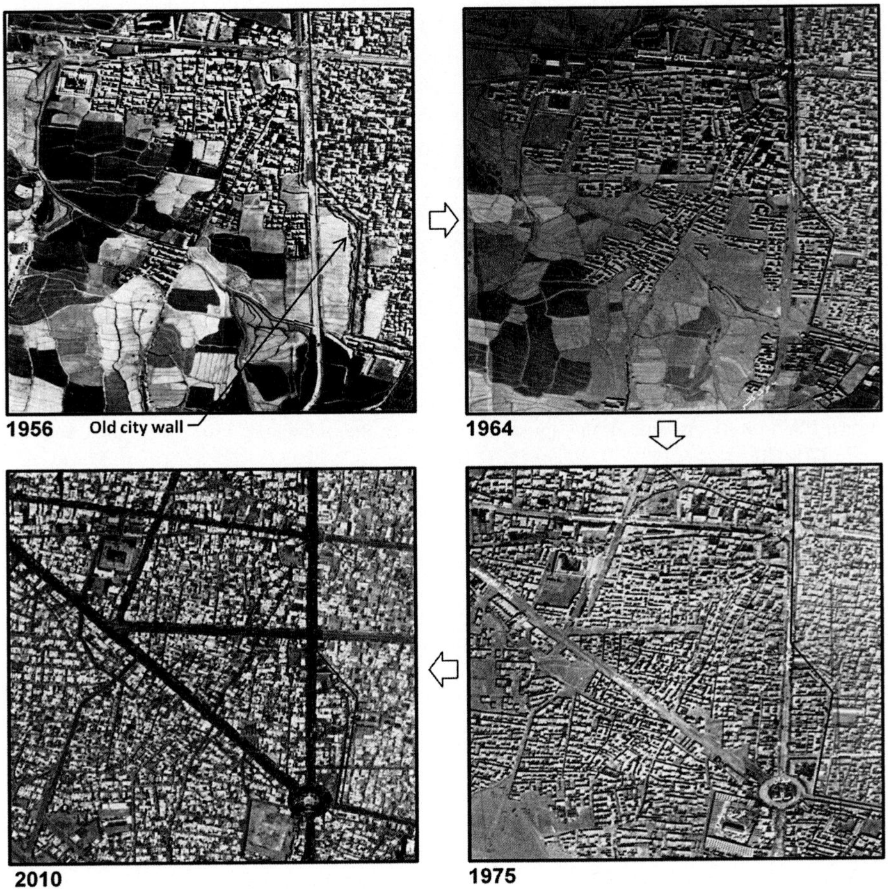

Fig. 2.6 The development and transformation process of the southwest part of Sabzevar. (Authors based on the map by Zanganeh (2003a, b, c) and Google Earth (2010). The base map reproduced by permission of Yaghoub Zanganeh)

2.8 Conclusion

The qanats as sustainable irrigation systems had a key role in the foundation and development of agricultural settlements on the Iranian plateau. The local water conditions and the possibility to build qanats determined the location of the main Iranian cities in this region, in the plains between the high mountains and the desert on the outer edge of the central plateau of Iran. The ancient Iranians constructed qanats and irrigation networks primarily for household water supply and farming, but at the same time these networks determined the shapes and growth patterns of the cities. In fact they created a platform for city formation and urban evolution. Thus the water supply system is one of the crucial influences on the morphology of the Iranian desert cities. For example, the urban pattern of old Sabzevar matches the irrigation network and the hierarchy of the old Sabzevar streets follows the hierarchy of the water distribution network. This network kept its role as a formative force in urban development even after losing its original function. The urban development pattern of Sabzevar is based on the transformation of farming lots to urban residential spaces. The disused irrigation network and the path system connected to it kept their shape and changed into streets and alleys; the dried farms were divided into smaller lots of land for residential buildings.

References

English PW (1968) The origin and spread of qanats in the old world. Proc Am Philos Soc 112 (3):170–181

English PW (1998) Qanats and lifeworlds in Iranian plateau villages. In: Transformations of Middle Eastern natural environments: legacies and lessons conference, Bulletin 103, Yale University, pp 187–205

Ghobadian V (2009) Sustainable traditional buildings of Iran, a climatic analysis. Islamic Azad University Press, Dubai

Google Earth. Imaginary Date: 8/4/2010. Sabzevar, 36°12′15.58″N, 57°40′27.78″E, elevation: 961 M. Available from http://www.google.com/earth/index.html (online). Accessed 12 July 2012

ICQHS (2009) An introduction to International Center on qanats and historic hydraulic structures. Available from http://www.icqhs.org/Files/PDFs/ActivityReport2008.pdf (online). Accessed 12 July 2012

ICQHS (2012) International Center on Qanats and Historic Hydraulic Structures, UNESCO Category II Center. Available from http://www.icqhs.org/English/subject.aspx?sbid=9a21f816-71ae-4220-8c53-4c4d32542d6a (online). Accessed 12 July 2012

Iran Hydrology (2005) Iranian qanat database. Available from http://www.iranhydrology.com/qanat/qanatlist.asp (online). Accessed 22 July 2012

Kheirabadi M (2000) Iranian cities: formation and development. Syracuse University Press, US

Nadji M (1973) The end of "Kariz's civilisation" in Iran, vol 22(9). Sokhan, Tehran, pp 938–945 (in Persian)

National Imagery and Mapping Agency (1999) Joint operation graphic Sabzevar, AIR series, sheet NJ 40-15, 1:250000. National Imagery and Mapping Agency, US

Sankakukei IK (2012) Blank map of the Iran. Available from http://english.freemap.jp/map.php?area=asia_e&country=iran (online). Accessed 12 July 2012

Semsar Yazdi AA (2010) Qanat, from practitioners' point of view (in Persian), Tehran. Water Resources Management Organisation Press, Iran

World of Maps (2012) Map of Iran topographic map. Available from http://www.worldofmaps.net/en/middle-east/map-iran/topographic-map-iran.htm (online). Accessed 12 July 2012

Zanganeh Y (2003a) Sabzevar photomosaic map based on 1956 aerial photos, sheet a1, 1:4500. Research Center of Geographic Sciences and Social Studies, Hakim Sabzevari University, Iran

Zanganeh Y (2003b) Sabzevar photomosaic map based on 1964 aerial photos, sheet a1, 1:5500. Research Center of Geographic Sciences and Social Studies, Hakim Sabzevari University, Iran

Zanganeh Y (2003c) Sabzevar photomosaic map based on 1975 aerial photos, sheet a1, 1:6500. Research Center of Geographic Sciences and Social Studies, Hakim Sabzevari University, Iran

Chapter 3
In the Shadow of Written History: An Exploration of the Tehran Bazaar as a Non-place

Farzaneh Haghighi

Abstract To answer the simple question 'what is the Tehran bazaar?', one will fall into an architectural and socio-political labyrinth of historical classifications, definitions and descriptions based on travellers' accounts, map productions, the monarchies' role in shaping cities, revolutions, theories of city evolution, the Iranian versus Islamic city controversies, archaeological approaches to civilisation and even linguistic approaches to Persian roots of the word 'bazaar'. The commonly accepted definition is that the Tehran bazaar is a linearly structured marketplace, and a united socio-cultural entity consisting of several public buildings of varying forms, functions, and historical values. This labyrinthine approach firstly ignores the transformative nature of the bazaar and presents it as a still, immobile and silent complex of static places; and secondly leaves little room to investigate the plurality of events and multiplicity of meanings taking place perpetually in this market place. This chapter utilises concepts of 'anthropological place' and 'non-place' by the French anthropologist, Marc Augé, in order to open new possibilities of looking at the multivalent context of the Tehran bazaar. It aims to go beyond the conventional understanding, to present an insight to a marketplace, problematising any enduring meaning.

Keywords The Tehran bazaar · Tehran urban planning · Marc Augé · Non-place · Anthropological place

3.1 Introduction

To answer the simple question: 'what is the Tehran bazaar?', one will fall into an architectural and socio-political labyrinth of historical classifications, definitions and descriptions based on travellers' accounts, map productions, the monarchies' role in shaping cities, revolutions, theories of city evolution, the Iranian versus

F. Haghighi (✉)
Faculty of Architecture, Design and Planning, The University of Sydney, Sydney, Australia
e-mail: farzane.haghighi@gmail.com

F.F. Arefian and S.H.I. Moeini (eds.), *Urban Change in Iran*,
The Urban Book Series, DOI 10.1007/978-3-319-26115-7_3

Islamic city controversies, archaeological approaches to civilisation and even linguistic approaches to Persian roots of the word ‘bazaar’. The commonly accepted definition is that the Tehran bazaar is a linearly structured marketplace, and a united socio-cultural entity consisting of several public buildings of varying forms, functions, and historical values.

This labyrinthine approach firstly ignores the transformative nature of the bazaar and presents it as a still, immobile and silent complex of static places; and secondly leaves little room to investigate the plurality of events and multiplicity of meanings taking place perpetually in this market place. In other words, we need to enter into this complexity without denying its incessant transformations.

This chapter utilises concepts of ‘anthropological place’ and ‘non-place’ by the French anthropologist, Marc Augé in his text *Non-Places; an Introduction to Supermodernity* (1995), in order to open new possibilities of looking at the multivalent context of the Tehran bazaar. It aims to go beyond the conventional understanding, to present an insight to a marketplace, problematising any enduring meaning.

3.2 The Tehran Bazaar in Socio-Political and Architectural Discourses

Following the decree of the Safavid king, Tahmasb I in 1553, the Tehran bazaar was built and ran as the heart of the city’s wholesale and retail activities (Madanipour 1998: 35). Affected by several urban planning schemes in the 20th century under Qajar, Pahlavi and Islamic governments, it transformed from a complex network into a distinguishable area demarcated by four streets: ‘15th of Khordad’ Street (previously called Buzarjomheri Street) on the north, ‘Mowlavi’ on the south, ‘Khayyam’ on the west, and ‘Mostafa Khomeini’ (previously Seerus Street) on the east. During the Qajar era, the earlier city walls were demolished by *Agha Muhammad Khan Qajar,* and replaced by an octagonal form with fifty eight spear-head shaped bastions as an imitation of the Parisian fortification pattern (Alemi 1985; Barthold 1984). This new setting did not last long. In the early 20th century, the second urban renewal programme transformed the fortified city into an open matrix, intertwined with modernisation and industrialisation programmes by Reza Shah, the founder of the Pahlavi monarchy. The arrival of motor vehicles and the regime’s desire to control urban populations were followed by construction of streets, squares and monuments during the Pahlavi monarchy and the Islamic government. The Tehran bazaar remained the main axis of the city and a residential neighbourhood during the Qajar era. However, it suffered dramatically by wide avenues cutting through later in the Pahlavi period. Under the Islamic government, the commercial and spatial centrality of the Tehran bazaar declined because of reasons such as the rise of the free trade zone and the shift of everyday interactions from the Tehran bazaar to the commercial streets and outskirts of the city (Keshavarzian 2007: 178–186).

The literature on the Iranian bazaar as a socio-economic concept is vast (Rotblat 1975; Ghandchi-Tehrani 1982; Larijani et al. 1993; Karimi 2008; Keshavarzian 2007; Bonine 1981; Thompson 1981; Ashraf 1981).[1] The Iranian bazaar as representing an urban social class and particularly as an economic institution resonates with the early writings of an Iranian-American sociologist Ahmad Ashraf in the 1970s, whose writings influenced several generations of sociology, economy and Iranian urban studies.[2] Adopting a Marxian-Weberian socio-economic methodology, he conceptualises the bazaar within social and urban hierarchies before and after the industrialisation of the 19th century in Iran (1969: 54). In his essay 'The Roots of Emerging Dual Class Structure in Nineteenth-Century Iran' published in 1981, he particularly explores the social formation of class before and after the nineteenth century. The pre-industrial Iranian society before the nineteenth century, Ashraf suggests, was composed of three fluid social hierarchies: the patrimonial strata (men of the pen, men of the sword, mainly the tribal leaders); the religious strata (men of sacred knowledge); and the strata of bazaar communities (men of trade, men of crafts, workers, peddlers, slaves, and beggars) (Ashraf 1981: 7). Such stratification in preindustrial society, Ashraf suggests, was plastic and relatively homogeneous because of three major reasons: first, the dominance of religion in all the social strata; second the physical proximity of various classes living in the same district; and third similarities in their life style) (Ashraf 1981: 17). Therefore, the citadel, the mosque and the bazaar corresponded to the urban hierarchy of the pre-industrialised homogeneous society.

In late 19th century, the Tehran bazaar was the focal point of wholesale trade. The middlemen arranged the relation between the village and the capital. The centrality of Tehran in Iran's economy in the late Pahlavi period concentrated several marketing networks 'under one roof'. In a historical perspective, a large body of literature has associated the bazaar with social movements, collective actions and crowd mobilisation in this period. Prominent theories have been raised by Ahmad Ashraf who outlined the relation between the mosque and the bazaar in urban upheavals, Nikkie Keddie who introduced the bazaar workers as an urban social class, Ervand Abrahamian who explored the bazaar as a traditional middle

[1]The only two books particularly on the Tehran bazaar are *Bazaar and State in Iran: The Politics of the Tehran Marketplace* written by Arang Keshavarzian published in 2007 in English and *Bazaar-e Tehran: Motaley-e Ensan Shenasi Eghtesadi* (The Tehran bazaar: An economic anthropological study) by Somayeh Karimi published in 2008 in Persian. Keshavarzian, states that the diverse fields of architecture, anthropology, economics, sociology, history, and political science have represented it as a place, an economy, a way of life, a class, and a symbol of Middle Eastern or Islamic life (Keshavarzian 2007, 40) and concludes four dominant conceptualising perspectives: the bazaar as a traditional type; a class; an informal economy; and, a product of informational scarcity (Keshavarzian 2007: 46–62).

[2]Naser Fakouhi, an Iranian cultural anthropologist suggests that Ashraf's contribution to Iranian urban sociology studies have remained one of the major resources in these fields (Fakouhi 2006: 334). For instance see the special issue of *Goft-o-gu* (*Dialogue*), a journal on culture and society published in 2004 in Persian on 'the bazaar', which trace the Ashraf's influence on the recent socio-economic studies in Iran (—, #776; Heidari 2004 #761; Keshavarzian 2007 #409; Piran 2007: #638).

class in relation with the crowd and finally, Arang Keshavarzian, an Iranian political economist at the Middle Eastern and Islamic Studies, New York University, who concluded that the current de-politicised and de-localised bazaar is incapable of mobilising the crowd. The bazaar's definition in these accounts is described as a response to an external event—state economic policies, revolutions, taxes, urban development and urban uprisings—rather than as an event itself.[3] However, Keshavarzian's approach stands as a different research. He sheds light on the social structure of the bazaar rather than its response to an external event.

The bazaar for Keshavarzian is 'a series of socially embedded networks within a bounded space that is the mechanism for the exchange of specific commodities' (2007: 16). As such, the Tehran bazaar is a network. A network bounded in a space; called an urban district; a distinguishable demarcated reign at the heart of the city; or, a collection of commercial narrow arteries. Although few socio-political studies investigate the bazaar as an event itself, their architectural accounts remain in the traditional morphological and functional spatial understanding. In sum, they borrow the same definition of the bazaar from architectural history (Keshavarzian 2009; Parsa and Keivani 2002). The bazaar here is a static linear structured marketplace, and a united socio-cultural entity consisting of several public static buildings varying in form, function, and historical value. This definition has not changed since 1898.

For instance, the American geographer Michael Bonine presents a definition of the bazaar less concerned with the urban context in which a marketplace operates, but rather focused on its constitutive elements as a complex structure in its own right and thus the bazaar becomes a complex of 'public institutions' (1989). Moreover, most commonly accepted formal analyses would suggest that the bazaar is divided into: *rāsta* (the main linear route with several shops on both sides); *raste* (the minor linear route, not as important as the main path); *dalan* (a covered alley that connects two *rāstas* or the interior and the exterior of the buildings. It also contains shops that sell goods of different kinds); *sara* (or khan, large complexes with a spacious central courtyard surrounded by one or two storey rows of rooms where goods could be stored and people lodged); *gaysariya* (a passageway with large doors); and *dokkan* (any unit of shops). According to bazaar's function, it is comprised of *ghahve khane* (coffee house); *chaikhane* (Tea house); *khorakpazkhane* (restaurant); *hammam* (bath house); and *mosques*(Pirnia 2003: 123–126).[4]

It is important to mention that there is no consensus on the classification of spaces within the bazaar. However, various shop types and pathways are the most recognised elements. Other components such as the mosque, the square, the

[3]For a comprehensive analysis of the theories of collective action on the Iranian revolution see: (Parsa 1988).

[4]This approach was also followed by Hossein Soltanzadeh, an Iranian architectural historian, from Tehran University, known for evaluating the contemporary city with the old city's principles (Habibi 2005, 56), who has particularly worked on the Iranian cities. He investigates the bazaar with particular emphasis on it as an object through historical, functional and spatial concepts. The difference between his approach to Pirina's is their debate on including square as one of the bazaar parts not as an independent urban space. This kind of debate is pervasive in bazaar morphological analyses.

bathhouse or the residential area have been discussed if they are a part of the city or the bazaar. In the functional schema, there emerges a common term that the bazaar is a 'complex of, centre of or collection of…'; with the rest of the definition being disputed by architectural historians. The understanding of the nature of the bazaar has become all the more ambiguous as there is no agreement on the territory of the bazaar's region, area, district or realm. Fortunately, recent historians of Iranian studies have taken this geographical ambiguity and complexity into account. This turning point has eloquently been reflected in the writings of Michael Bonine in *Is there A Middle East?* (2012a) and Firoozeh Kashani-Sabet in *Frontier Fictions* (1999). Although the Tehran bazaar is not the main concern for these two scholars, their approach is valuable in acknowledging the confusing inconsistency of maps in historical knowledge production and map deployment in land-based nationalism (Bonine 2012a: 92). Bonine and Kashani-Sabet have opened the discourse of territorial conflicts in current wars in the Middle East.

In 2012, the Tehran bazaar no longer consists of a residential neighbourhood; it no longer operates in political opposition against the government, it is no longer a place for authentic merchants and it is no longer a centralised commercial centre for import and export of Persian carpets. It has transformed into a battlefield between the Tehran municipality, the Iranian Heritage Cultural Organisation (ICHO) and the *bazaaris* (the merchants in the bazaar). The battle suggests clearly that this place has not yet completely drained of its eventfulness. As such we might bring to the fore the necessity of a framework that enables us to improve the architectural understanding of the bazaar in the plurality of events, the multiplicity of meanings, and the diversity of the crowd.

3.3 'Anthropological Place' and 'Non-place' by Marc Augé

Marc Augé, the French anthropologist trained in the 1960s ethnologists' school of thought, was drawn into cultural anthropology and creative fiction by Claude Lévi-Strauss, a French structuralist anthropologist. By the mid-1960s Augé made several trips to West Africa studying the culture of this post-colonial society. The concept of *non-lieu* or 'non-place' which appeared in his writings is derived from the field work in the Ivory Coast and study in Paris and became the subject of his book titled *Non-Places: Introduction to an Anthropology of Supermodernity* originally published in 1992 in French and translated into English in 1995.[5]

The notion of non-place—also found in forms of nonsite, nonloci or commonplace—as the subject of critical emphasis on territory has a pivotal importance in French philosophy post May 1968; parallels the political turn from modernity's

[5]In French: *Introduction á une Anthropologie de la Surmodernité*. In 2008 the second edition with a new introduction titled *Non-Places: An Introduction to Supermodernity* was published in English.

questions of time and history to postmodernity's question of space and geography. These concepts have emerged in the work of Michel de Certeau, as a negative quality of place, in Michel Foucault's writings underpinning the vacancy at the 'heart' of modern anthropology and archaeology, and in Gilles Deleuze's analysis of the gap between the visible and the sayable. However, Bruno Bosteels, Professor of Romance Studies at Cornell University, argues that Augé stands as the 'reference point' for this idea among French theorists (2006: 118).

For Augé, the non-place is the production of supermodernity with practical implications in particular concrete places of circulation, consumption and communication, such as highways, airports, supermarkets, clinics, refugee camps, hotels, sport clubs, meeting rooms, parks and auditoriums (Augé 2008: viii). The clear image of these contemporary places, has given way to several architectural debates, deploying Augé's theoretical proposition, for instance in the reading of England's M1 motorways, central New Delhi market or the Polish socialist housing projects (Merriman 2004; Favero 2003; Svensson 2012). What remains less explored is the coexistence of non-place within an anthropological context. In other words, Augé defines non-place not just as the birth place of postmodernity, but also in close relation with the 'common past and the contradictory interpretation of events' (Bosteels 2006: 119; Augé 1995: 73).

Augé conceptualises the 'anthropological place' within an ethnological experience of the indigenous Alladian country in West Africa. The study starts from the locality of a murder through interrogation of a cadaver and then goes beyond an African society to provide a model for investigating contemporary French public spaces. The initial foundation, letting the two different cultures establish a dialogue is the 'geometric' essence of place in non-traditional anthropology. The geometrical structure of itinerary, intersection and centre, as Augé suggests, is the characteristics of anthropological place where people want them to be 'places of identity, of relations and of history' (1995: 52). For the relation between identity and place, he raises the simple example of birth. The phenomenon of being born is a spatial event in a way that each person will be assigned to a residence and the actual place of birth will constitute the individual identity. Augé continues that this place is not a distinct and solitary concept, since the birthplace is a shared territory formed by the relations others have established to it before. From the moment that identity combines with relation by a minimal stability, he suggests, places will become 'historical'. To approach such a historical place, an anthropologist's intellectual role is ambiguous in a way that s/he is not in search of a 'paradise', but rather to conduct careful spatial and geometric observations (1995: 54–56). This kind of observation investigates the lines (routes), intersection of lines (crossroads), and points of intersection (centres) in human geography which are political and temporal in nature, therefore, can be applied to contemporary places.

France, as Augé argues, usually known as a centralised country composed of towns with their own monumental centres, presents the complexity of an anthropological place simultaneously being geometrical—in city's spatial arrangements such as town hall, churches, markets, cafes and monumental streets—and historical (1995: 69). The 'illusion' to the past, taking place in contemporary French is represented in street names, spatial landmarks and more importantly in metro stations. For Augé, metro, a supermodern space, is a non-place or the opposite of utopia. In the midst of the vague zone of an anthropological place and non-place, Augé suggests that 'if a place can be defined as relational, historical and concerned with identity, then a space which cannot be defined as relational, or historical, or concerned with identity will be a non-place' (Augé 1995: 77–78). At first glance it seems that Augé has drawn an apparent and distinctive line between the two spatial concepts; however, the complexity of identity, relation and history does not construct two clearly defined zones. In his terms, place and non-place are two 'opposed polarities'—one will never disappear and the other will never finish. Augé states that they are in constant conversation, as a 'palimpsest on which the scrambled game of identity and relation is ceaselessly rewritten' (1995: 79).

The anthropological place and non-place coexist by distinct definitions but not necessarily distinct territories, rather mobile and overlapping geometric reigns. The interplay of identity, relation and history has established the shared linkage between the two spatial concepts. The description of the non-place is not just correlated to this shared zone, but also in Augé's terms, it is the real measure of the supermodern era. Accelerated by an 'overabundance of events, spatial overabundance, the individualisation of references' supermodernity has just come into living reality, and yet needs spatial rethinking (1995: 40). The non-place is unprecedented because of the unpredictability of the overabundance of events taking place in it and it is not solely a pure form but is also composed of the individuals' relations with their surroundings. This domain, the non-place, daily receives large numbers of individuals or, in Augé's terms, the 'average man'. The isolated and solitary passenger who is fabricated by a space invaded with signboards, screens and posters that offer instruction for use. They face the same instruction, voices and messages as others do. Augé suggests that texts and words govern the inhabitants of this banal utopia, and as such their identity is not formed through unformulated and complex local references but is based on a shared identity of relatively temporary users (1995: 101). Big supermarkets stand as a clear example where customers of the retail space are guided by signs and labels, involved in a 'silent dialogue'. The warning signs in highways, identity cards in airports and price labels in duty-free shops, as Augé states, designate the solitary person with their contractual identity. Here emerges the most critical aspect of life in the supermodern era where people confront with the same image of themselves as millions of others and history is reduced to a spectacle.

3.4 The Tehran Bazaar in the Shared Territory of Anthropological Place and Non-place

As discussed, the bazaar has been conceptualised as a linearly structured marketplace; and as a united socio-cultural entity consisting of several public buildings that vary in form, function and historical value. Built in the 16th century, it was affected by several urban plans in the 20th century under Qajar, Pahlavi and the Islamic governments. Today it is known as a fixed central urban district demarcated by distinguishable linear north-south and east-west streets. This chapter aims to situate the Tehran bazaar in the shared ambiguous territory of the two conceptualisations of place. Firstly, by introducing the bazaar as an anthropological place where people want it to be a place of identity, of relations and of history, and then by raising the transformed relation of the people to that place. This suggests that the Tehran bazaar is also a non-place challenging the enduring meaning imposed upon it.

In terms of identity, the Tehran bazaar was one of the oldest living quarters in the city of Tehran (Gurney 1992: 56). Referring to the 18th century maps of Tehran, the town planner and academic Ali Madanipour, describes the spatial structure of the bazaar as the main axis of the clear functional structured city. In other words, this marketplace was not an isolated commercial area, but rather immersed in the city's life at least during the Qajar era. Urban studies have benefitted greatly from several travellers' accounts by providing an historical picture of the Tehran bazaar.[6] The multiple functions of this marketplace allow historians to delineate the identity that the bazaar provides for merchants, known as *bazaari*, participating in trade within this environment. The *bazaaris*' strong sense of community and cooperative activity has resulted in abundant socio-political studies analysing the social structure of this urban population (Ghandchi-Tehrani 1982; Ashraf 1988; Ashraf and Abrahamian 1983; Mozaffari 1991).

The *Asar* magazine, the journal of the ICHO published a special issue in July 1980 in which the Tehran bazaar is introduced as in urban decay and in danger of fire and destruction. This special issue can be regarded as one of the first attempts raising the importance of saving this historical centre. The comprehensive report by Bavand Consultants in 2008 set the aim of its project as regenerating this old district. Nagsh-e Piravash Consultancy carried out a study on the old fabric of Tehran. The project aimed to restore the historic heart of the capital. In 2012, the Tehran municipality published a report titled the 'Socio-Cultural Evaluation of the Tehran Bazaar' in which the bazaar became the representation of the identity of people living in Tehran. The restoration project on the Tehran bazaar, started in 2005 is divided between three stakeholders: Tehran Municipality; the consulting engineers, and the general contractor. The bazaar district has been broken up between four consultant engineers carrying out projects from façade restoration, arching passageways and squares, roofing, entrance construction and pavement renovation.[7]

---

[6]For more information on travelers accounts see: (Yarshater 2001). Mostly cited observations by viceroy of India and British foreign secretary, statesman, traveler and writer George Nathaniel Curzon (1859–1925, visited Persia from Sep. 1899 to Jan. 1910) describes the vaulted Tehran bazaar as a succession of low brick domes) (Curzon 1892 #446@330). The same approach can be found in the writings of British archaeologist, traveler, historian and spy, Gertrude Margaret Lowthian Bell (1868–1926, who spent six months during 1892 in Persia), and the British oriental scholar Edward Granville Browne (1862–1926, who visited Persia in 1887 for a year).

[7]*'Tarh, Tafakor, Memari', 'Chehelsotoon'; 'Mana'; and 'Sharan'* are the four architectural consultants involved in the Bazaar's restoration project.

These institutions share the perspective that the Tehran bazaar as the centre of the city has to represent the traditional/historical identity of Tehran. Such a geometrical argument emerges at the intersection of being authentic and being at the centre simultaneously. In Augé's terms the anthropological place is geometric, not just in simple defined spatial forms, but rather in the complexity of passing activity along routes, gatherings and meetings in crossroads, and religious or political embodiment in centres (1995: 59). In this respect, the Tehran bazaar is a centralised urban territory, corresponding to identify, relation and history.

Today we understand the bazaar as the sum and rcsult of predictable events in the past: industrialisation, modernisation, revolution and globalisation. This narration resonates with Augé's outstanding interpretation that 'the revolutionary event (and in this sense the revolution is exemplary as an event) cannot be reduced to the sum of the factors that make it possible and, after the event, understandable. We would be quite wrong to limit this analysis to the case of the Revolution alone' (1995: 27–28). History cannot explain the shift in our understanding from a functional definition to statistical data. History, Augé tells us, 'is on our heels, following us like our shadows, like death. History meaning a series of events recognised as events by large number of people … events we believe will count in the eyes of future historians' (1995: 27).

In contrast to the notion of bazaar's decline, a survey conducted in 2005 suggests that the Tehran bazaar is the largest centre of attraction in the city (Yousefi Far 2010: 17), where every day 400,000 people pass through. Of all its commercial units, 41 % act at a national level and there is no other commercial centre in Tehran capable of competing with the bazaar because it has become the main link between producers and consumers in the whole country (2010: 50). Although the overabundance of unpredictable events such as assassination, drug smuggling, hiding illegal refugees and cargo loading night life are not documented in historical account of the bazaar, they can be found in newspaper reports.[8] Most importantly, the anthropological place resists accepting the excess of experience. Since all of these unexpected events take place in the same historical location of an anthropological place, again this supports the claim that the Tehran bazaar is also a non-place. It is relational, contingent and in perpetual process. The Tehran bazaar, in the interplay between two spatial concepts of anthropological place and non-place, points to their shared permeable and ambiguous zone with no defined, but recognisable frontiers. In Augé's terms' in the concrete reality of today's world, places and spaces, places and non-places intertwine and tangle together. The possibility of non-places is never absent from any place. Place becomes a refuge to the habitué of non-places' (Augé 1995: 107).

The renovation project in the bazaar is complicating contemporary relation to the history; however, incessant references to the street with transforming names, religious signage and metro stations are structuring our relation to the non-place.

[8]The assassination of Asadollah Lajevardi, the warden of Evin Prison in 1998 in the Tehran bazaar. The Seyed Ismaiel bazarche in well known for second hand commodities and smugglers.

On the '15 Khordad Street' we find ourselves in a pedestrian avenue which was previously called Bouzarjomheri Street. General Karim Bouzarjomehri, was the mayor of Tehran in the 1930s, in charge of the destruction of the city wall during Reza Shah's modernisation programme (Madanipour 1998: 37). However, '15 Khordad' stands for the '5th June 1963' known as the starting point of the Islamic revolution of 1979, when the demonstration started from the bazaar in response to the Shah's arrest of Ayatollah Khomeini. Oscillating between the recent 1930s and 1980s, pedestrians encounter several entrance signs to the bazaar's routes by the names which no longer exist. A brown metal sign points to the 'tobacco' bazaar, but in that route you can find anything but tobacco. This particular name resonates with the Tobacco Rebellion of 1891–92, when the Shi's clergy issued the fatwa on banning tobacco against the Qajar politically driven economic concessions to the West.

To distinguish the threshold between the Tehran bazaar as an anthropological place and as a non-place, there is a need to rephrase the question of defining boundaries to the manner in which our relation to that zone is directed and regulated. This is what the word 'non-places' designates: 'clearly the word "non-places" designates two complementary but distinct realities: spaces formed in relation to certain ends (transport, transit, commerce, leisure), and the relations that individuals have with these spaces.' (Augé 1995: 94).

3.5 Conclusion

This chapter demonstrated the limitations of the architectural and socio-political conventional understanding of the bazaar as a complex of static places. To go beyond formal and functional approaches of available studies, it raised the necessity of a perspective enabling us to capture the transformative essence of the bazaar. I utilised Marc Augé's theories of 'anthropological place' and 'non-place' and discussed their shared ambiguous territory in which the Tehran bazaar may be redefined not just as an historical place, but rather as a non-place challenging the enduring meaning imposed upon it.

To understand these new meanings, the chapter does not aim to find the 'real' meaning of the bazaar, but rather suggests an *oscillation* between two entangled geometrical conceptions of anthropological place and non-place in which there is room for an overabundance of events taking place and generating new meanings. There is, therefore, no surprise there are several discussions about whether it is alive or dead, whether it is historic or modern, whether it is the commercial centre of Tehran or the economic hub of Iran and whether it is the centre of the city or a city within a city. We might accept that the Tehran bazaar is not just an immobile entity or a fixed urban district. It is defined and redefined ceaselessly by the relationships associated with it.

References

Alemi M (1985) The 1981 map of Tehran: two cities, two cores, two culture. Environ Des: J Islamic Environ Des Res Centre 1

Ashraf A (1969) Historical obstacles to the development of a bourgeoisie in Iran. Iran Stud 2(3)

Ashraf A (1981) The roots of emerging dual class structure in nineteenth-century Iran. Iran Stud 14 (2)

Ashraf A (1988) Bazaar-mosque alliance: the social basis of revolts and revolutions. Int J Polit Cult Soc 1(4)

Ashraf A, Abrahamian E (1983) Ahmad Ashraf: bazaar and mosque in Iran's revolution. MERIP Rep 113(113)

Augé M (1995) Non-places: introduction to an anthropology of supermodernity (trans: Howe J). Verso, New York

Augé M (2008) Non-places: an introduction to supermodernity (trans: Howe J). Verso, New York

Barthold W (1984) An historical geography of Iran (trans: Soucek S). Issawi C, Lewis B (eds) Modern classics in near eastern studies. Princeton University Press, Princeton, New Jersey

Bonine M (1981) Shop and shopkeepers: dynamics of an Iranian provincial bazaar. In: Bonine M, Keddie N (eds) Modern Iran: the dialectics of continuity and change. State University of New York Press, Albany

Bonine M (2012a) Of maps and regions: where is the Middle East? In: Bonine M, Amanat A, Ezekiel Gasper M (eds) Is there a Middle East?: the evolution of a geopolitical concept. Stanford University Press, Stanford CA

Bonine M (2012b) Bazar [Encyclopaedia Iranica online edition]. Center for Iranian Studies, Columbia University 1989. Available from http://www.iranicaonline.org/articles/bazar-i. Accessed May 2012

Bosteels B (2006) Nonplaces: an anecdoted topography of contemporary French theory. Diacritics 33(3)

Curzon G (1892) Persia and the persian question. Longmans, Greens & Co, London

Fakouhi N (2006) Ensanshenasi ye shahri (Urban Anthropology). Nashr-e ney, Tehran

Favero P (2003) Phantasms in a "starry" place: space and identification in a central New Delhi market. Cult Anthropol 18(4)

Ghandchi-Tehrani D (1982) Bazaaris and clergy: socio-economic origins of radicalism and revolution in Iran. Ph.D. Thesis, Department of Sociology, The City University of New York, New York

Gurney J (1992) The transformation of Tehran in the later nineteenth century. In: Hourcade B, Adle S (eds) Téhéran, Capitale Bicentenaire (Tehran, paytaht-e devist-sale-ye Iran). Paris - Téhéran: Institut Français de Recherche en Iran; Louvain, Belgique: Diffusion, Editions Peeters

Habibi SM (2005) Intellectual trends in the contemporary Iranian architecture and urbanism (1973–2003). Daftar pajoohesh haye farhangi, Tehran

Heidari A (2004) Asnaf va dowlat. Goft-o-gu 41

Karimi S (2008) Bazaar-e Tehran: Motaley-e ensan shenasi eghtesadi. In: Hassanzade A (ed) The Tehran bazaar: an economic anthropological study. Afkar, Tehran

Kashani-Sabet F (1999) Frontier fictions: shaping the Iranian nation, 1804–1946. Princeton University Press, Princeton, New Jersey

Keshavarzian A (2007) Bazaar and state in Iran: the politics of the Tehran marketplace. Cambridge University Press, Cambridge, UK

Keshavarzian A (2009) Regime loyalty and bazari, representation under the Islamic Republic of Iran: dilemmas of the Society of Islamic Coalition. Int J Middle East Stud 41(2)

Larijani HH, Tohid LN, Nasiripour M (1993) Ketab name-ye bazaar (The Bazaar Bibliography). Bonyad-e dayeratolmaaref eslami and sazman miras-e farhangi, Tehran

Madanipour A (1998) Tehran: the making of a metropolis. Wiley, New York

Merriman P (2004) Driving places: Marc Augé, non-places, and the geographies of England's M1 motorway. Theory Cult Soc 21(4–5)

Mozaffari M (1991) Why the bazaar rebels. J Peace Res 28(4)

Parsa M (1988) Theories of collective action and the Iranian revolution. Sociol Forum 3(1)

Parsa A, Keivani R (2002) The Hormuz Corridor: building a cross-border region between Iran and the UAE. In: Sassen S (ed) Global networks, linked cities. Routledge, London

Piran P (2007) Nazariye rahbord va siyasat-e sarzamini jame'e Iran: pazhohesh-e shahr shenasi. Andisheh Iranshahr 2(6)

Pirnia MK (2003) Ashnaiee Ba Memari-Ye Eslami-Ye Iran (An Introduction to the Islamic Architecture of Iran). Daneshgah-e elm va sanaat-e Iran, Tehran

Rotblat HJ (1975) Social organization and development in an Iranian provincial bazaar. Econ Dev Cult Chang 23(2)

Svensson M (2012) The socialist housing projects as non-places in post-2000 Polish literary and cinematic narratives. East Eur Polit Soc 26(3)

Thompson CT (1981) Petty traders in Iran. In: Bonine ME, Keddie NR (eds) Modern Iran: the dialectics of continuity and change. State University of New York Press, Albany

Yarshater E (2001) The Qajar era in the mirror of time. Iran Stud 34(1–4)

Yousefi Far S (2010) Sargozasht-e bazaar-e bozorg-e Tehran, bazaar ha va bazaarche hay-e piramoni ye an dar devist sal-e akhir. In: Habibi H (ed) Chronicle of the Tehran grand bazaar: the bazaars and surrounding bazaarche since the recent 200 years. Tehran: Bonyad-e Iran Shenasi

Chapter 4
Re-visiting Three Neighbourhoods of Modern Tehran: Chaharsad-Dastgah, Narmak and Nazi-Abad

Rana Habibi, Bruno De Meulder and Seyed Mohsen Habibi

Abstract In 1945 the municipality of Tehran planned for the first time the development of large-scale Residential Neighbourhood projects. Over-population, increasing rents and land prices in central Tehran made de-centralisation and expansion necessary. With the approval of the first seven-year development plan in 1948, the construction of 'Low-Cost Housing' got a prominent place on the development agenda (Planning Organisation, implementation report of second socio-economy development, 1964). The 'Rahni Bank' [Mortgage Bank] was appointed by the Tehran Municipality as the executive organisation for low-cost housing projects. The organisation included young Iranian architects with European training and new ideas for the creation of a modern Tehran and adhering to the credo of modernist urbanism. The first modernist residential neighbourhood 'Chaharsad-Dastgah' was built in 1946 for government employees. As in so many (official) discourses (worldwide) of the time a lot of emphasis goes to efforts to attain affordable prices, what was supposedly related to materials, economy of scale, modern techniques and middle-class as a target group. In 1952, the new cabinet approved the construction of two large townships, respectively, Narmak in the north-east and Nazi-Abad fields in the south of Tehran. In Narmak and Nazi-Abad designs, the challenges of modern and traditional life style were clearly seen. As such, their development histories (conception, implementation, reception and appropriation) are crucial to understand the metamorphosis of Tehran from a

R. Habibi (✉) · B. De Meulder
Department of Architecture, Urbanism and Planning, University of Leuven, Leuven, Belgium
e-mail: rana.habibi@asro.kuleuven.be

B. De Meulder
e-mail: bruno.demeulder@asro.kuleuven.be

S.M. Habibi
Faculty of Urbanism, Tehran University, Tehran, Iran
e-mail: smhabibi@ut.ac.ir

F.F. Arefian and S.H.I. Moeini (eds.), *Urban Change in Iran*,
The Urban Book Series, DOI 10.1007/978-3-319-26115-7_4

traditional city into a modern metropolis. The present chapter attempts to shed light on these development histories as a particular case study that articulates the specificity of Iranian modernism and modernisation.

Keywords Iranian modernism · New towns · Residential neighbourhoods · Tehran

Abbreviation

AIAD The Association of Iranian Architects-Diploma

4.1 Introduction: Development and Modern Questions

> In recent years the breeze of 'modernism' has transformed Iranian social life and has created a 'Spirit of Modernity' in people—a spirit that is perfectly manifested in architecture. Soon after, the Iranians encountered modern problems and their survival. The era pushed people to make more efforts and its impacts would be seen in all aspects of social life. We have responsibility towards future generations and we should shoulder our responsibility in the best possible way; if we stick to our traditions and consequently fail to take our responsibility, we prove to be worthless, meaning that we are unable to take care of our progress and excellence and we fail to have an adequate understanding of the 'Spirit of the Time'. (Hovanesian 1946)

Between 1945 and 1953 Iran underwent major socio-urban changes. It witnessed the fast formation of urban middle classes, development of technocratic parties, establishment of modern urban planning and so many other aspects of modern cultural and urban development. The Association of Iranian Architects-Diploma (AIAD) [in Persian: Anjoman-e-Architect-ha-ye-Irani-ye-Diplome], created in February 1944, was one of the outcomes of this transition period. It was formed by 38 young architects under presidency of Kayghobad Zafar and vice presidency of Ali Sadegh. Other members included Abbas Adjdari, Manouchehr Khoursand, Naser Badie, Vartan Hovanesian and Iraj Moshiri. These Europe-trained architects returned to the country with a mission: re-constructing, re-thinking and questioning the current situation of Iranian cities. The 'regrettable' situation of Iranian cities was unacceptable for them and they assigned themselves the duty to find suitable solutions (Moshiri 1946). The AIAD advised Tehran Municipality and introduced international principles of building, contacted foreign cultural associations and attended international conferences.[1] They opened an architecture library (mainly with donations from Italy and France), organised exhibitions (e.g. on English architecture) published journals, etc. In short, the introduction of modern architecture and urbanism was on the agenda.

[1]In 1948 Ali Sadegh and Iraj Moshiri from AIAD and Iraj Shams from Tehran Municipality attended to International Union of Architect (UIA) congress in Lausanne, Swiss.

The journal ‘Architect’, started in 1946 by these young architects demonstrates their missionary mindset: ‘detoxify Iranian cities, diagnosing the weakness of Iranian cities, and discuss the best solutions’ (Moshiri 1946). As self-declared technocrats, and in line with the modernist discourses and practices in Europe, they criticised the traditional practices and theoretical approaches of Iranian urbanism. Mass housing was a main focus of that period (post-war) in Iran (like most countries that were witnessing large demographic growth). Population statistics saw the population of Tehran grow rapidly from 362,000 in 1921 to 1,205,000 in 1941 (Habibi and Hourcade 2005). Housing became scarce. As in Europe and elsewhere[2] architects produced and reproduced their typical discourses ‘social’, ‘public’, ‘mass’ and collective housing.

This significant growth of Tehran had many reasons: surely until the coup of September 1941, the State focused solely on the development of Tehran (while neglecting other cities), Tehran absorbed the largest share of immigration because of new factories, the expanding governmental sectors, the establishment of university and collages in Tehran, and on top of all, the monopoly of foreign trade given to the capital. It encouraged traders from other cities to migrate to Tehran (Adjdari 1946a, b). During the Allied occupation of Iran (after September 1941), lots of people moved to Tehran for security reasons (Adjdari 1946a, b). All these make the post-war ‘housing crisis’ (Adjdari 1946a, b) in Tehran understandable. Housing and land prices increased dramatically while speculators exploited the country’s financial situation. As a result governments faced high housing demands while supply remained low (Adjdari 1946a, b). Over-population was the result. Therefore, (like in so many countries in this ‘development decades’) massive ‘low-cost housing’ projects became the solution. The housing crisis was evidently intertwined with relating questions of urban growth. Architects faced questions like ‘how to develop cities quickly?’, ‘what is the fastest way to expand cities?’, ‘which housing typologies are suitable for modern Iranian citizens?’, and ‘how does development remain affordable?’

4.2 From Practice to Planning: The Case of Chaharsad-Datgah and the Seven-Year Planning

> Although the steady industrialisation of the country made the number of workers increase day by day, nobody would think of their welfare. So many children would die every day in those chaotic, contemptible and dirty places, lacking suitable light and fresh air; these unhealthy conditions would gradually and cruelly break the spirits of these young people and would penetrate into and poison all their cells. It is a gradual death, it is a crime! … The government had to think of a solution for this strange disease called ‘housing famine’ which

[2]For example the CIAM conference on minimum housing.

had afflicted the workers and low-income families. For this reason, architects and urbanists have to accept the responsibility to tirelessly make efforts to address the housing problem and to develop a 'lively and dynamic' city which required the establishment of modern neighbourhoods with all sorts of facilities. (Hovanesian 1947: 25)

4.2.1 Chaharsad-Dastgah (400 Units): The First Modern Low-cost Housing Project in Tehran

Residents governmental employees
Housing units 400
Area 60,000 m^2
Architects Ali Sadegh; Manouchehr Khoursand; Hossein Sadegh; Abbas Adjdari [AIAD members]
Executive Rahni (Mortgage) Bank
Investor Melli (National) Bank

Ideas concerning low-cost housing in the form of new neighbourhoods developed gradually. They were first formulated with the establishment of a 'Construction Firm' in 1936 and afterwards 'Beh-Saz' Firm in 1938 by Mohammad Ali Sheibani (an architect that knew about French low-cost housing and urban development through discussions on 'Louis Loucheur'[3] on a conference in Switzerland) with the support of Ali Akbar Davar, then functioning as the Minister of Justice. Davar himself was familiar with the ideas concerning low-cost housing from the time he was in Paris. Sheibani returned to Iran after his studies in France and started to redact his writings about 'low-cost mass housing plans' and 'the creation of new neighbourhoods'. These studies and actions were interrupted at the start of the World War II (Sheibani 1946).

After the war, in order to repair war damages and deal with housing shortage, European governments invested massively in new neighbourhoods and public housing. Iranian architects and urbanists who studied abroad and observed this 'progress' there, envisioned similar developments in Iran after returning to the country in response to what they called the 'dysfunctional' situation of cities. Abbas Adjdari, a member of the AIAD, commented: 'thinking and studies are just a small part of the process. Catching financial investments such as governmental loans and getting support from banks are crucial. Otherwise there will be no move towards the action phase.' (Adjdari 1947).

Mohammad Ali Sheibani also, in a small note in the *Architect Journal*, (1946: 28) labelled low-cost housing as a national movement:

> Today is the time to start building small, beautiful housings for the virtuous homeless government employees. In developed countries, construction of housing is an important

[3]Labor minister of France from 1 June 1928 till 2 March 1930.

> part of government planning in the post-war period. In Iran, we should follow the idea of housing productions like other countries in the world. For the development of the country, we should not be waiting for government assistance. In our opinion, as architects and urbanists, and despite problems, we should start developing actions to construct low-cost housing; we indeed hope to have a beautiful Tehran in the near future. By constructing new neighbourhoods and housing according to modern urbanism principles, we hope to create a spirit of health and joy. Is the basis of all social reform in progressive countries not housing and nothing else?

In 1945, the construction of low-cost housing appeared on the agenda of the prime minister Saham-Soltan Bayat's cabinet. Several commissions were formed in 'Rahni Bank' [Mortgage Bank] and 'Melli Bank' [National Bank] to study the technical and financial possibilities for the mass production of low-cost housings and new neighbourhoods. With the change of the government, the subject was dropped. Nevertheless, the next cabinet of Ghavam-Saltaneh (1946) defended the necessity of low-cost housing for low-income government employees in parliament. The municipality of Tehran prepared a plan for 400 low-cost housing units for Tehran. Mehdi Mashayekhi, the Mayor of Tehran calculated a budget of 50 million Rials for this important experiment. Finances were provided by the 'Melli Bank' and the Tehran municipality. The 'Rahni Bank' took responsibility for the construction (Adjdari 1947). The municipality proposed 'moat lands'[4] [khandagh in Persian] at the eastside of Tehran (Shahnaz Street), but the moat was 3 to 4 m deep. The supervision committee of Rahni Bank saw three options: first using an apartment typology (this would make selling difficult in terms of division and ownership)[5]; second, filling the moat (which was quite costly). The third (and finally withheld) option was switching to another site. The supervisory committee bought 130,000 m^2 of land in the southeast end of Tehran (south of 'Doshan-Tappeh' road and in front of the 'arms factory') and 60,000 m^2 to the south of the '500-bed hospital' was also bought for construction of the new neighbourhood 'Chaharsad-Dastgah'. Thus a 400-unit neighbourhood for low-income employees was constructed with the investment of Melli Bank and a design team appointed by Rahni Bank. All designers were AIAD members. They worked under the supervision of Ali Sadegh, Manouchehr Khoursand, Hossein Sadegh and Abbas Adjdari (Fig. 4.1).

Justifying the typology, Abbas Adjdari, the spokesman of the group explained: 'Low-cost housings in other countries in the world are apartments, but in Iran, given the climate and local culture, apartments are still not acceptable. Also in legal terms, the ownership of an individual apartment is still undefined. Therefore, it was decided to construct a complex of single houses for this assignment of low-cost housing.'

---

[4]Ditch lands were part of municipality's properties.

[5]Only in 1965 law of apartment possession was approved by the parliament, before that possession of land has recognition.

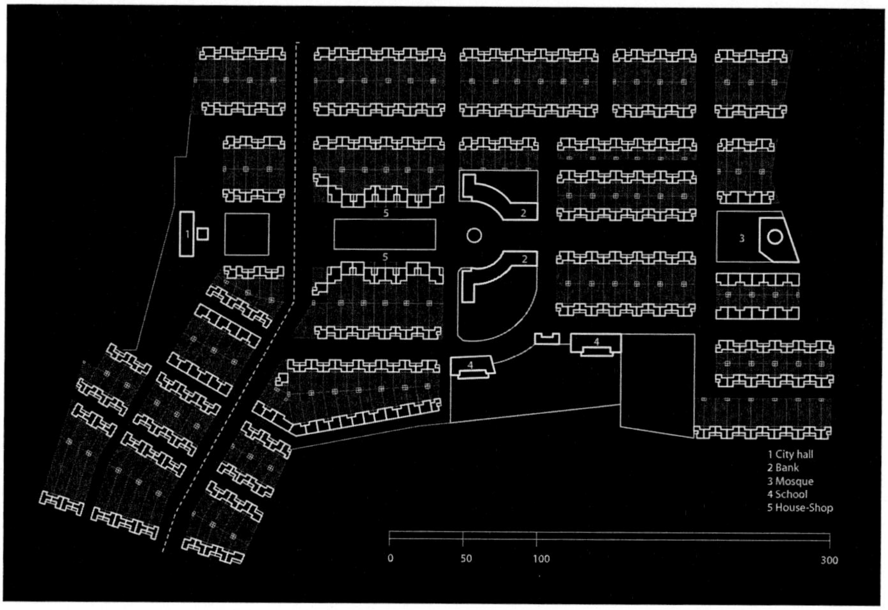

Fig. 4.1 Schematic presentation of Chaharsad-Dastgah (Habibi 2015, base map from National Cartographic Center with permission)

According to Adjdari and Khoursand, the main principles of housing complex 'Chaharsad-Datgah' are:

- A maximum use of sunshine. So, most of the houses have an orientation towards a north-south axis guaranteeing the entry of suitable south light.

On average, the area of each plot (buildings with their individual yards) is 170 m^2. Housing plots have access to the street from the north as well as from the south. House plans were designed based on access to streets. In this way all housing units have a room facing south. They receive south light either from individual yards, or from the street. All housing had enough light in line with health principles in urbanism (Adjdari 1947).

- Providing large green spaces inside main squares of neighbourhoods (as a compensation for the minimum space of the housing units: minimum space for housing and maximum public space; minimum for individual spaces, maximum for collective spaces).

The geometrical shape of new neighbourhood was a rectangle of 420 by 260 m. The main square located on the north side of neighbourhood had a length of 165 and width of 45 m. It provided a large green public space for the dense neighbourhood (Khoursand 1947).

- Dark and narrow streets were avoided. A large vast green boulevard (which was also suitable for busses) was provided instead.

Main streets had a width of 20 m and sub-streets had widths ranging from 8 to 15 m. East-west axis is an internal street blocked with two public buildings: the municipality building in the west and the mosque at the top end of the east.

- Centralising commercial and administrative activities in the main neighbourhood square, equally accessible for all residents.

All shops were located around the main neighbourhood square. A covered gallery in concrete gave access and simultaneously protected the stores. The colonnade of the gallery brought light in and generated air circulation, making a healthy place (in contrast with old bazaars which were totally covered and just had small roof openings to get air and light). These colonnades also create a new concept of bazaar by consideration of modern urbanism principles (Adjdari 1947). On the eastern side, inside the semicircular form, on top part a police office was located while Melli Bank stood beside it and on the opposite part a post and telegraph office and Rahni Bank were sited. To the south of the neighbourhood a state elementary school was constructed with 12 classrooms and a large amphitheatre for the neighbourhood. On the west of the school a land was provided for a hospital and the Ministry of Health. Also, deep wells and a water supply and treatment resources with all accessories were located on the west side of the neighbourhood behind municipality building. Laundry location and public bathroom (Hammam) stood in the same area.

- By providing row housings with green gardens, a green structure was generated in the neighbourhood.

Economy was of utmost importance (as exemplified in granting 20-year loans). Decorations were deliberately omitted. Four main housing typologies (with two, three, four and five rooms[6]) were designed. Terraced housing combined spatial economy with ease of construction. Flat insulated roof was chosen for houses, with plain brickwork to façades and plain white finish to interiors. Kitchens, storages, independent yards with a trough, and balconies were common elements in all types. Also in all cases a staircase was continuing to the roof-terrace (the sleeping space during summer time). The only exceptions were the houses facing the neighbourhood's main square constructed in two floors, with the ground floor belonging to shops and houses above them with a variety of typologies of two, three and five rooms. Toilets in all types were provided outdoors in yards and connected with shallow wells which work with septic tanks. These wells were 70 m higher than the level of underground waters for protecting drinking water from waste water (Khoursand 1947). Composite beam and vault systems were used to construct floors and roofs.

[6]Rooms traditionally in Iran had different functions during day and night. Used as living and dining rooms during days and sleeping space during nights.

Two years after the start of 'Chaharsad-Datgah', the 'First Seven-Year Development Plan of Iran' was drafted and a 'plan and budget council' was established. The first seven-year plan coincided with the Marshall Plan of 1948: the American Programme (1948: 52) through which the European economy was reconfigured according to American models that emphasised continued productivity gains. Technically speaking, oil fuelled this economic growth path. Hence, the world economy's oil dependency became structural. Consequently, the Middle East moved to the centre of foreign-policy strategies for a number of Western nations, becoming also a site of ideological conflict between the Soviet Union and the United States, with both nations acting out their political differences through technical aid and development projects as well as cultural exports (Isenstadt and Rizivi 2008). In this context, the first seven-year plan provided a framework for social, economic and cultural development, following the model of 'modern societies' such as the U.S. and Western Europe. The main goal of the first seven-year plan was raising the standards of living by, among other things, promoting knowledge and improving lives and livelihoods. The policy was implemented through direct government intervention and the creation of institutions for the construction of 'Low-Income Housing', financed by long-term loans.[7] For this reason, in 1951, the 'Planning Organisation' and 'Bank-e-Sakhtemani' [Construction Bank] were established through the investment of the Agriculture Ministry and Iran Insurance Company, under the supervision of Doctor Mohammad Mosaddeq, the then prime minister. Following the first seven-year plan, the transfer of ownership was approved in the parliament of the 'dead lands'[8] from the state to the 'Bank-e-Sakhtemani'. They were reserved for constructing 'Low-Income Housing' for middle classes. 'Low-Income Housing' was included as a subsection of social affairs in the urban-law plan Local and global architectural ideas were used for the conception of one, two and three room single-family houses.[9] Space was provided for service centres and public facilities and all this contributed to the definition of the 'new modern neighbourhoods'. The 'low-cost housing' plan was an answer to the high housing demand (itself resulting from the growing 'urban population') specifically increasing the amount of available housing for rent. This is particularly important when the housing supply is limited.

These years-long experiences with housing, from the analysis of housing problems till the practical delivery of modern housing for the new low-middle classes were part and parcel of the seven-year planning that in general also brought international expertise into the country.

---

[7]Planning Organisation report of implementation of First Seven-Year plan, 1964, Part I, the Municipal Civil Actions.

[8]According to law, article 27, dead lands are the arid, anhydrous and dry and out of use which are not usable for agriculture. The concept of dead lands has religious roots: according to Islam and religious books, non-agricultural lands should consider as public benefits and only if private owner revive dead land as housing or orchards can have ownership of the land (Nasrollah 1955).

[9]Before approval of the law of 'Apartment Possession' in 1965, Iran had Land Possession Law, which means most of estate housing before 1965 was made as single house typology.

4.3 The Practice of Modernism: The Case of Narmak and Nazi-Abad

> Once a wasteland with no residents, Narmak that was quickly changing into a beautiful modern city … Narmak is now a perfect example of Iranian architectural taste … This city has all modern life requirements, cinema, theatre, hospital, playground, water and electricity and so on … Department stores similar to those in European cities and large boulevards have given this modern town urban-life qualities. Keyhan daily newspaper, (Bank-e-Sakhtemani Journal, 1955: 36)
>
> Low-cost housings of Narmak and Nazi-Abad were built in seven months without the Government's support and by prepayment from land purchases. Small housings with two, three and four rooms were constructed according to the latest style of European housings. Low-cost but beautiful and liveable housing is a good example for private investors to learn how to invest their money in construction of low-cost housings. The Visit of Dr. Tarchand, Ambassador of India, to Narmak New Town, (Bank-e-Sakhtemani Journal, 1955: 36)
>
> Today our Iranian-European lifestyle makes a big problem for the design of our housings. Even our knowledgeable and leading architects cannot respond to the complex demands of clients; people want a perfect house with a minimum space and the high quality of traditional housings and at the same time with consideration of all the elements of modern life; for instance, when a family asks an architect to design a house for them, they may say: 'we want grand living and dining rooms and private rooms for ourselves; also we want a bathroom with a warm floor and a bath like French Salle de Bain.' Thus, to design a house, we need to consider the disciplines and traditions of the owner; architects, therefore, need to analyse traditional Iranian housings and include important elements of traditional housing like climate, as well as religious and local materials in their designs. (Pirnia 1955: 6)

4.3.1 Narmak, New City for Middle-Classes

Assigned housing plots areas 224,683,479 m^2
Public space areas 184,632,422 m^2
Governmental institutes areas 97,405,328 m^2
Assigned housing units numbers 7500 units
Housing units' areas 200–500 m^2
Population 25,000

With the approval of the law on the 'Registration of Dead Lands' in 1952, large fields around Tehran (beyond its boundaries) were possessed by the State. This allowed the cabinet of Doctor Mosaddeq to approve the construction of two large towns (housing complexes). In both sites, Nazi-Abad fields (in the south) and Narmak (in the northeast of Tehran) the 'regulations of urbanism' would be observed (Habibi et al. 2011). The 'Bank-e-Sakhtemani'—established in 1953 with the future development of Tehran as its target—was funded by Agriculture Ministry and Tehran Municipality. These two projects, especially Narmak, became a model of Iranian modernist low-cost housing and characterised the 'modern Iranian neighbourhood'. Narmak was designed by the 'Bank-e-Sakhtemani' design team that was comprised of members of the AIAD. Naser Badiee headed the design team while Abbas Adjdari was one of the main designers. The following elements are specific for Namrak new city (Fig. 4.2):

- Public spaces: re-articulation of the traditional neighbourhood centre and grid

The entire area of Narmak new city was 600 ha. This equals a sixth of the total area of Tehran at that time. Of this amount 184,632,422 m^2 were allocated to public space and streets, and 97,405,328 m^2 to public institutions. The rest, 224,683, 479 m^2, was allocated to housing, 36 % of which designated as open space. This included squares, public spaces and streets. Each neighbourhood has a small square, used as children's playground and green space. Branching from every square were four eight-metre wide dead-end streets, car-free zones made safe for residents. Other straight and main streets were allocated to cars. Climate considerations led to

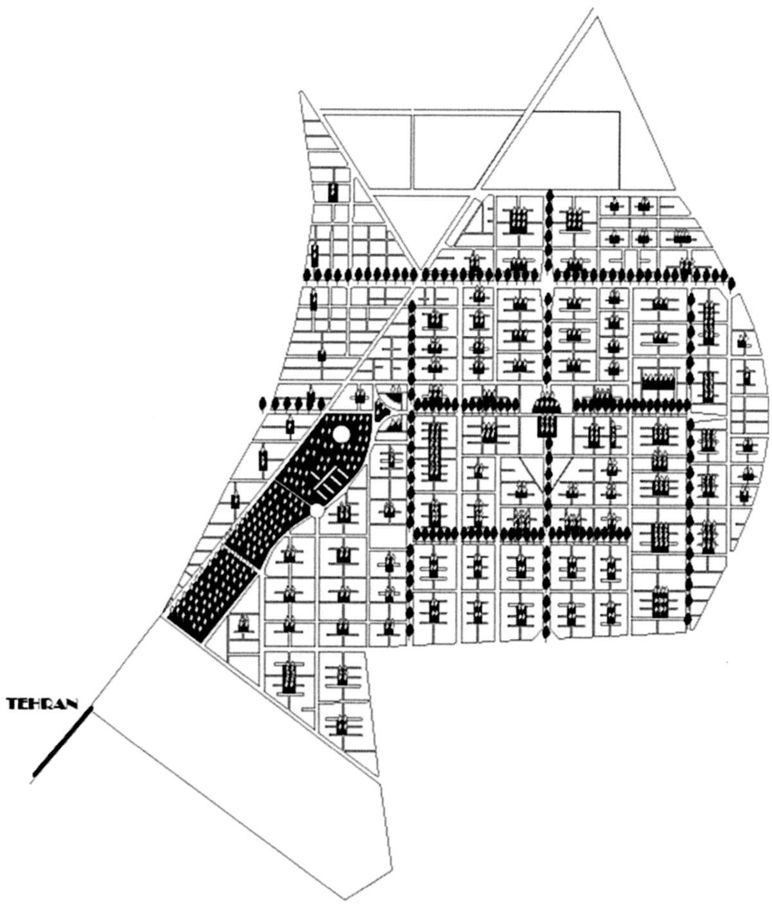

Fig. 4.2 Schematic presentation of Narmak grid, boulevards and common gardens (Habibi 2015, base map from National Cartographic Center with permission)

the north-south orientation of buildings. A grid organises the plan. Streets were designed in a rigid way without any flexibility in form like European and American cities (Adjdari 1956). The east-west axis crosses the central square and ends in the east with the town's Capital Mosque. To its west the city park ends the axis. The park to the west and on the windway, contributes to the desirable climate of the town. In the middle of each neighbourhood, a small square with garden was provided as a gathering and leisure space for families. It simultaneously prevents heavy traffic inside the neighbourhoods. Only residents were allowed to enter by car.

- Public facilities: combination of modern and traditional activities

In the proposal several public facilities were defined for Narmak, though not all of them realised. Public institutions, cinema, theatre, sports hall, mosque, urban park, hospital, orphanage, school and shopping stores form the core of urban facilities were proposed for Narmak New City. There was also a modern art museum which was never realised. It was to have an exhibition hall for 200 painting panels and 500 sculptures, and also an amphitheatre for 500. Located in a 150 m^2, it was to be in five levels connected by ramps in the middle of each floor (Haj-Dayi 1955).

Both sides of the east-west axis were allocated to shopping and commercial activities. The electricity centre and small industrial activities were provided in the northern part to protect housing located south of industrial activities from industrial pollution. The strategy to decentralise public facilities meant that each neighbourhood had its own small supermarket, school and public bathroom (Khodayar 1955a, b).

In terms of urban infrastructure, before implementing housing units, the 'Bank-e-Sakhtemani' provided the water supply network for the neighbourhood. To the north-east, in 'Shian' forest, a well was drilled and its water directed to an aqueduct that became the main water source for the neighbourhood. For the first time in Tehran, reinforced concrete was used in the aqueduct. This way, the amount of water consumption in summer and winter became manageable and the underground water table could keep reserves for dryer times. Beside the water path, lines of trees were planted. A total of 20,000 trees were planted in the neighbourhood (Khodayar 1955a, b).

- Housing costs and typology

Architects used straight streets around housing plots and assembled them in a grid. Each plot measures between 200 to 500 m^2. Rectangularly shaped and with north-south direction, they correspond with traditional land division rules, making them suitable for the semi-arid climate of Tehran. The master plan of Narmak foresaw 7500 housing plots. The length to width ratio of plots is two to one, which allows an adequate space for the build-up area and a vast open space for each plot that guarantees each house suitable light and air. Roofs were built with reinforced concrete. All buildings were equipped with water pipes and electricity supply.

Bahmani[10] bricks were used for façades. Roof fences and window frames were realised in white concrete. Since the site had been transferred from the state to Bank-e-Sakhtemani, the bank charged buyers only the building costs—40 Rials per metre, while prices elsewhere in Tehran had already reached 400 to 500 Rials (Moeenfar 1955). Part of construction materials were provided for contractors by the bank. The price of buildings excluding water storage, was 1150 Rials per square metre. The price of water storages were 850 Rials per square metre. For two-room housing, a deposit of 1500 Rials was asked from applicants. The total price for two-room housing was 95,000 Rials (Moeenfar 1955).

Two-room units:

- Area: 54 m^2 for building +balcony, 12 m^2 area for water storage, in total 66 m^2
- One 3 × 4.70 m room,
- One 3 × 3 m room,
- Kitchen, storage, balcony and water storage are provided for each unit as well.

Three-room units:

- Area: 61 m^2 for building without balcony, 11 m^2 area for water storage, in total 73 m^2

In total the price of a two-room house was 90 % of the price of a three-room house.

The bank also constructed 50 large four-room houses, 10 of which for its employees in the centre of the town and the rest in the west side, and sold them through lottery. A five-room unit was also constructed at the centre, but people did not like the type. Meanwhile, people asked Bank-e-Sakhtemani to construct housing with balconies so 46 units with four rooms and balconies were constructed (Sarafian 1960).

- Housing factory: KALAD pre-fabrication system; one week one house

The French prefabrication system manufacturing KALAD Factory was established in Narmak. The factory functioned 250 days per year and had a production capacity of 250–300 four-room, 80 m^2 units. The factory also produced prefabricated toilets and kitchens. This semi-automatic prefabrication system made sandwich panels based on 1.10-m modules with maximum spans of 4.40 m. One housing unit could be assembled every day. The system had sound and heat insulation and double-leaf walls providing suitable protection against earthquakes (Khodayar 1957).

Discussions on interwoven issues of construction quality, financing, etc. led to an understanding required to combat homelessness among architects (Khodayar 1955a, b):

[10]Popular bricks belonging to 40s, 50s.

- To avoid the assignment of plots to people
- To increase the level of finance provided by institutions dealing with housing (bank investment)
- To rationalise production and hence decrease the price of building materials
- To classify housing according to typologies and numbers of rooms, area and finance them accordingly
- To define a reasonable maximum possible term for housing price repayment
- Housing construction should go hand in hand with the simultaneous provision of basic infrastructure like roads, water and electricity
- Construction of low-income housing shouldn't be limited to Tehran, concluded the architects of the time. Other cities should provide low-income housing. This strategy would, they argued, avoid immigration from other cities to Tehran.

4.3.2 Nazi-Abad, New Workers Town

Workers city for southern side of Tehran
Entire areas 2,150,000 m^2
Public space areas 1,000,000 m^2
Assigned housing units numbers 2000 units
Housing units' areas 200–650 m^2

Historically, Nazi-Abad was a small agricultural village in the vicinity of Tehran. The enormous urban development of Tehran and the massive industrialisation transformed Nazi-Abad willy-nilly into a small city. In 1953 a modern neighbourhood of Nazi-Abad was constructed for workers of the railway factory. On the north side the land was adjacent to the slaughterhouse of Tehran, on the east to Ali-Abad lands, on the south to Yakhchi-Abad and on the west to Javadiyeh region and 'Ghale Morghi' airport (one of Tehran's military airports).

- Land distribution

The entire area of Nazi-Abad was about 250 ha and was mainly allocated to housing and public institutions. An important third part was allocated to streets and public spaces. The Nazi-Abad lands were divided into 2700 housing plots. Plot areas ranged from 150 to 650 m^2. After only 2 years in 1955, 400 families lived in this new town. Among these plots, Bank-e-Sakhtemani constructed 80 housing units for sale and 65 housing units for the relocation of villagers (labelled slum dwellers at the time). In total, 10 ha were allocated to public spaces, among others, in the form of small neighbourhood gardens (as in Narmak new city) (Maleki 1955).

- Housing typology

The main housing types are two- and three-room houses. Bank-e-Sakhtemani, divided the site into plots of 200–650 m^2 and distributed it between families. The bank itself constructed 16 low-cost houses with two rooms (49.48 m^2 each, costing

45,000 Rials and requiring deposits of 25,000 Rials), or three-rooms (7300 m^2, 70,000 Rials with deposits of 35,000 Rials) and four-rooms: (9000 m^2, costing 90,000 Rials, with deposits of 45,000 Rials). Candidates for a house in Nazi-Abad had the opportunity to visit housing at Narmak and ordered their units as matching with them (Moeenfar 1955).

- Public facilities

To avoid the unpleasant smell of the slaughterhouse, a 10-ha area was set aside for forestation as an isolator in the north side. Water supply came from the old aqueduct, north-east of the neighbourhood. For electricity generation for the new Nazi-Abad neighbourhood, the Alstom Power Station was established. The Bank also financed the construction of four markets and two schools, and also established two factories for mosaic and concrete shafts and allocated 203,056 m^2 to a communications building.

The experiences in Nazi-Abad show a lot of small shortcomings—heavily debated at the time[11]—which all in all are not so surprising, given the fact that we are witnessing the first practices of modern urbanism in Tehran (and Iran).

4.4 Conclusion

To conclude we can briefly highlight two major debates as follows:

First, Chaharsad-Datgah, Narmak and Nazi-Abad exemplify urban development in Iran in the immediate post war period. Low-cost housing was coming up not only as responding to the needs of the middle classes but also as shaping of future metropolitan of Tehran, which by having 1,6 million residents at the end of 1956 was in the beginning of its great expansion. These three cases were prototypes of eastern and southern neighbourhoods of Tehran. Farah-Abad and Nirou-ye-Havayi neighbourhoods were constructed with the same logic of Chaharsad-Datgah, Tehranpars was based on the same logic as Narmak, and Yakhchi-Abad in the south was influenced by Nazi-Abad, which means they partly shaped Tehran's urban growth. During this decade one witnessed also the start of growth in other cities in Iran [like construction of low-cost housing in Isfahan (the Industrial City) and Ahvaz (the Oil City)]. On the other hand these three cases showcase the formation of the 'New City' and urban transformation. Also in terms of architecture, these three projects show the major shift in housing typologies (low-rise housing for single families instead of multi-room housing for group families). Neighbourhood formation is rational, almost mathematical and hierarchical, not organic: multiplication of houses and then public spaces on different scale levels. It is not that the

[11]Lack of supervision from Bank-e-Sakhtemani and failure in accurate and precise implementation of housings (wrong material in places, mistakes in limits of lands and building location towards streets) and other technical problems like displacement of staircases to the roof, miscalculation of wall heights and shortening of roofs (Sarafian 1960: 18).

lay-outs of streets, squares, etc. have formal intentions embedded in them. They simply result from the carving out of the city mass with the aim to create access. They are grids generated by a rationale of access. The grid is not an aim in itself. By and large, the modern idioms lying behind these projects are massive scaling, rational multiplication, the setting up by the state of a system with its components ranging from financial institutions (banks, etc.) to construction, delivery mechanisms and infrastructure.

Second, these three low-cost housing quarters were good examples of localising modern urbanism by architects with totally different approaches from those of 1930s top-down modernist (nationalist) architecture in Iran. Thinking of climate and culture, and above all people's customs, and trying to combine modern and traditional elements in low-cost housings (new towns) are all signs of a vernacular modernism shaped in the early 1950s and followed until the 1970s (for instance in Shushtar New Town by Kamran Diba). However, by the beginning of 1960s the mainstream forms of development shifted towards Western modernism leaving behind the approaches of 1950s architects who tried to define Iranian modernity and re-read traditional concepts in a modern way.

The emergence of the metropolis and the development of new cities, together with the emergence of Iranian modernity or vernacular modernism (in other words localisation of modernism) are the two significant developments of the 1945–1953 period: a period we can call the period of 'practising Iranian modernism'.

References

Adjdari A (1946a) Housing problems in Tehran and other Iranian cities. Architect J 1(1):15–16
Adjdari A (1946b) Housing problems in Tehran and other Iranian cities. Architect J 2(1):51–52
Adjdari A (1947) Construction of low-cost housings in Iran. Architect J 4(1):125–133
Adjdari A (1956) Elements in Narmak and Nazi-Abad proposal. Bank-e-Sakhtemani J 4(1):21–22
Habibi R (2015) Tehran Middle Class Housing 1945–1979, Unpublished Ph.D. thesis, University of Leuven
Habibi M, Ahari Z, Emami R (2011) From demolishing fortifications to thoughts of highways: history of urban design in Tehran from 1930 till 1966. Soffeh J 50(20):51–61
Habibi M, Hourcade B (2005) Atlas of Tehran metropolis. Pardazesh va Barnamerizi-e Shahri Press, Tehran
Haj-Dayi A (1955) Narmak Museum. Bank-e-Sakhtemani J 2(1):28
Hovanesian V (1946) Architecture problems in Iran. Architect J 1(1):4–9
Hovanesian V (1947) Architecture problems in Iran II. Architect J 2(1):24–26
Isenstadt S, Rizivi K (2008) Modernism and Middle East: architecture and politics in the twentieth century. University of Washington Press, Seattle and London
Khoursand M (1946) Association of Iranian architects. Architect J 1(1):3
Khoursand M (1947) 400 units low-cost housings. Architect J 4(1):134–135
Khodayar M (1955a) How Narmak was made. Bank-e-Sakhtemani J 2(1):7–9
Khodayar M (1955b) Division of lands is not a solution for homelessness. Bank-e-Sakhtemani J 2 (1):21–23
Khodayar M (1957) The first 5 rooms house made by Kalad system. Bank-e-Sakhtemani J 6 (1):11–15

Maleki M (1955) Nazi-Abad. Bank-e-Sakhtemani J 2(1):39–40
Moeenfar M (1955) Housing brawl? Bank-e-Sakhtemani J 2(1):31–34
Moshiri I (1946) Our goal. Architect J 1(1):1–2
Nasrollah K (1955) J Bank-e-Sakhtemani 1(1)
Pirnia M-K (1955) Traditional Iranian housing. Bank-e-Sakhtemani J 2(1):6–8
Sarafian M (1960) Overview of housings. Bank-e-Sakhtemani J 2(2):16–24
Sheibani MA (1946) Appearance of low-cost housings in Iran. Architect J 1(1):28

Chapter 5
A Review of Urban Images of Tehran in the Iranian Post-revolution Cinema

Hamed Goharipour

Abstract Both metropolis and cinema are products of modernity. Since the end of the nineteenth century, the fortunes of cinema and the city have been inextricably linked on a number of levels, and it can be said that cinema, space and place need each other for self-description. First images of cities in the movies almost coincided with the birth of cinema and after a little time, all major cities in the world were shown in movies so that perhaps the social and physical changes in cities such as Berlin, Paris, Rome and cities of the United States can be reviewed through cinema along with other sources. This association was so strong that a new genre was born in the cinema: the 'city film' or 'urban symphony'. Subjects such as alienation, population growth, mass storage, informal settlements, urban semiotics etc. repeatedly and in different time periods have been displayed in movies. An analysis of a filmic representation of a city begins with observing how individual films represent the conditions of that city or its neighbourhoods in a specific historic moment, and then moves beyond seeing the film as mere representation of social reality to focus on how the cinematic text constructs and comments on those conditions. Any analysis of a city film can begin with an examination of the primary urban spaces portrayed in the film. How are spaces staged and linked or not linked to other spaces in the city? How are interior and exterior spaces, public and private spaces, defined? How do spaces reflect the social reality of race, class, and gender? In the recent decades Tehran has been the major location of Iranian films. The city has been analysed, sometimes as a context for the story and sometimes as the main theme of the film. The purpose of this chapter is a review of urban images of Tehran using Iranian films since 1980s until today. Accordingly, some movies are selected based on their themes of presenting urban problems and their relationship with

H. Goharipour (✉)
Department of Urban and Regional Planning,
Allameh Tabataba'i University, Tehran, Iran
e-mail: Hamed_goharipour@yahoo.com

F.F. Arefian and S.H.I. Moeini (eds.), *Urban Change in Iran*,
The Urban Book Series, DOI 10.1007/978-3-319-26115-7_5

Tehran, and then, the Tehran cityscape is reviewed through their perspectives. The chapter does this by focusing on the most important urban concerns reflected in movies.

Keywords Cinema and city · Tehran · Film · Tehran image

5.1 Introduction

City is an expanded, clear and also very complex entity. Theorists have long tried to present a comprehensive definition of the city, but all these definitions at best just describe some features of the city from a specific angle. In fact, from Plato's time (and even before) to now, everyone has described the city according to personal views and criteria. So, there are many definitions of the city in different texts occasionally contradicting one another.

Even though it is probable that cities deeply differ in terms of their raison d'être and culture or economy, it is clear that when anyone talks about the city, he or she has images and phenomena in mind which are different from those of rural or any non-urban areas.

Cities have provided many facilities for their residents, but they have also caused problems, which is why scholars, thinkers and artists in different fields have been led to think about the city, describe it, and also use their own devices to communicate their thoughts. This include cinema which, as an important media since the 20th century, has played a significant role in the representation of the city.

5.2 Cities, Urbanisation and Its Problems

It is clear that urbanisation has improved different aspects of human life, but its waves have also had other effects on new and former residents of urban areas, which have motivated theorists, thinkers, politicians and also artists to think and present ways to make urban life better. This process has more disadvantages in many developing countries. According to Fakouhi (2004: 40) 'the third urbanisation wave, developed more rapidly than developed countries, started after the second world war in many countries in the world'. It is in reaction to these difficulties that thinkers and theorists have described their proposed utopia.

Plato introduced a utopia whose first instruction was the subordination of philosophy. In the sixteenth century, Sir Thomas Moore published his famous book *Utopia* and described the ideal social conditions in an imaginary island. In the middle of the industrial revolution, some theorists who were renown predecessors of planning, most importantly Charles Fourier and Etienne Cabet, proposed their utopia realised in an ideal living complex.

Notwithstanding details of these utopias, what matters most is their reflections on urban problems as well as their proposals for solutions. In the meantime, the most common tools for them to publish their ideas were writing and in some cases sketching. They tried to use mass media to spread their theories and, as McQuail (2009: 21) has said 'mass media are one of resources of the power. It means that they are tools for control, management and innovation in society which could be a succession for power or other resource'. Thus, the cinema as a global and attractive medium has a special place here.

5.3 Cinema and the Other Arts

According to Encyclopaedia of Urban Studies, 'despite all the changes movie houses have gone through, the cinema has always been a vital element of urban life' (Hutchison 2010: 137). It is not easy to consider an accurate time as the moment of invention of cinema. It could be said that cinema invented and changed from an amusing job to a global industry during three decades before the World War I. The rapid advances of the cinema at the end of the 19th century and the beginning of the 20th, made many experts in different science and art disciplines realise its role and importance.

In the beginning, there was no tendency in other arts to recognise cinema as a new art; but cinema gave them an opportunity to communicate their experience with millions of viewers around the world in a new language. In the meantime, one is unlikely to find two arts disciplines as closely aware of one another as cinema and architecture. Their relationship started when cinema needed to use architecture for designing locations. Even though this usage of architecture was very simple in the beginning, because of the importance of the mise en scène, filmmakers thought more accurately about methods of using architecture even making changes favour of architecture. Their innovative use of architecture forced architects to think about cinema and some of them reviewed their past ideas. Extensive issues about common aspects of cinema and architecture led experts of these two areas to hold two pioneering conferences: CIAM and CICI.

5.4 City and Cinema

What is the relationship between the city and the cinema? The city and the cinema have long had a relationship starting with the appearance of the first images of cities in the cinema. Although cinema can be said to have flourished in Paris, Paris was not the only important city in its development. Technological and artistic developments were occurring in London, Berlin, Moscow and New York, all of which contributing in early improvements of filmmaking industries. So, according to Mennel (2008: 86), 'the growth of cinema was intimately tied to the growth of cities … capital for production was to be found in cities, and more profit could be made by locating movie houses there because the urban population had ever more expendable income and leisure time'. All these led to cinema recognising the city as its context.

Many authors have highlighted the effects of cinema on the city, society and culture, and urban studies have paid a special attention to films as a cultural perspective of cities, because, as Mark Shiel and Tony Fitzmaurice say (2001: 3), cinema is the 'innovative spatial form of culture'. John Rennie Short explains in his *Urban Theory* (2006) that modernity, capitalism and post-modernity link film studies to urban studies. Serge Daney, the famous film critic, has an attractive point in one of his articles: 'cinema is an inseparable part of city. It is not after or before it. It is one soul in two bodies' (quoted by Croot 1997: 75). So, the relationship between the city and the cinema is undeniable, but unfortunately there is a limited literature on this relationship especially about Iranian cinema and city.

5.5 Methodology

As mentioned, this chapter investigates the image of the city in the Iranian cinema. There are many methods for reading a film subject to one's purpose. The most common way of reviewing films is quantitative analysis, but it seems that this method is not appropriate for checking out the image of the city in the perspective of the cinema, and, nor is the presentation of conclusions by charts.

Instead, constructive analysis is used here to investigate the image of the city in cinema, which is a qualitative method aiming to extract the structure of a phenomenon. So, we will first concentrate on the general image of the city, and then the relationship between urban elements and their hidden messages.

5.6 Images of Tehran in the Iranian Post-Revolution Cinema

Undoubtedly, the most influential event after the revolution was the eight-year 'imposed war': a war whose impacts were not limited to frontlines, but also in rural and urban areas. This period has become the theme of many films some of which representing the situation of cities in the wartime, called 'City in War' here. *Vasle-e-Nikan* (Hatamikia 1991) is one of them.

The general image of Tehran in this movie is one of a quiet city turned to the location of a rocket war. In a few seconds the calm and quiet city is filled with ambulances and fire-fighting alarms. Many long shots from a city sky where rockets are moving, invokes the sense of fear. All streets are dark and dialogues about leaving the city add fuel to the mental space of fear.

In most of the war years and after that, big cities welcomed immigrants who had fled their homes due to the enemy's attack. These war refugees along with other immigrants, who came to Tehran and other major cities to find work, achieve greater prosperity etc. were forced in many cases to confront and deal with problems that sometimes made them to displace. Many post-revolution directors chose migration to big cities and its consequences as their themes and represented the city from the viewpoint of these immigrants. This city-image is named here 'City of Immigrants'.

Canary Yellow (Bani-Etemad Canary yellow 1988) narrates the life of a man called Nasrollah, who is fooled by an urban crook and loses all his money. He comes to Tehran to put his life back in order. But the urban man who Nasrollah takes refuge in is himself a crook, symbolising urban settlers who cheat poor people and steal their money.

The theme of migration to big cities is seen in many other movies. *Leaning on the Wind* (Farhang 2000) is the tale of a young man who hopes to gain wealth and have a better life. He leaves his wife for a while and moves to Tehran. But he is caught in the city by a woman, accused of murder and sentenced to death. *Mirage* (Tamjidi 1986) also narrates the story of a non-urban man's life whose dream to live in Tehran comes true, but he faces many problems, including house renting and living costs. So, he is forced to work overtime and also as a taxi driver. This pressure on him and his family, eventually convince him to go back home.

A Night in Tehran (Kazemi 2008) is another film, a comedy, taking on problems of migration to Tehran by a fantasy narrative. The film is the story of a young couple going to Tehran for a job interview. Before reaching Tehran, the couple's dreams of the future is interesting: a happier life in Tehran, excellent schools for children, great food, unique restaurants, a house near a park, etc. However, they arrive in Tehran after their adventures in a rainy night, facing unexpected things contrasting their dreams: a flood leaving a lot of waste on the streets, hotels where all their rooms are full, a crook who steals their possessions, etc.

Reasons for these representations of problems facing immigrants in big cities are not clear, possibly a sense of responsibility to prevent further migration. But statistics show that this common movie themes is not in line with realities, as almost all major cities in Iran have been facing significant population growth and the resulting urban sprawl. In fact, many of the problems and shortcomings of the existing cities are related to population growth. The growth in these years has caused problems such as marginalisation, poor economy, addiction and many others.

The other issue, and undoubtedly one of prime concern to anyone, is shelter. Non-standard construction, uncontrolled expansion of the city, shortage of services and infrastructure, housing and mental health problems resulting from poor housing were issues addressed in some Iranian movies. This city-image can be called the 'City of Horizontal Expansion'. *The Tenants* (Mehrjui 1986) may be one of the best examples of highlighting the housing crisis in those days (Fig. 5.1).

In the beginning of the film, Tehran is viewed from south to north by moving cars and people, and the camera finally stops in front of a building in a marginal

Fig. 5.1 Marginal areas and foothills of Tehran where now is being constructed on *The Tenants* (Mehrjui 1986. Reproduced by permission of Dariush Mehrjui)

area. Dialogues about the locations of buildings and construction of roads, bridges and parks leading to a long shot from foothills of Tehran suggest the city is being torn. In one sequence, with a pigeon sitting on a water tank on the roof, the tank moves and causes serious wobbles in the building. This beautiful scene depicts the instability of these buildings. *The Tenants* is a statement against non-standard construction that took place in Tehran since the 80s. Poorly engineered buildings, unprepared land, ownership disputes, and profiteering games of business owners and dealers are some of the problems which are represented in this movie.

On the other hand, population growth has resulted in the loss of old streets, the construction of highways, and the vertical, as well as horizontal growth of the city. So, a new cinematic city was born. This city-image can be called the 'City of Towers'. *Soltan* (Kimiaei 1996) addressed these issues. In one famous sequence of the film, Soltan, who is wounded by land speculators, speaks about his memories while he is standing in the middle of an urban freeway (Fig. 5.2):

> Our house was here (*pointing to a part of the highway*). Two hundred (*square*) meters of courtyard with five bedrooms, two gardens and a pond in the middle of it.

Then goes to another part of the highway and continues:

> Here was my room. There was my mother's room where her samovar was always on in it, Always! Here, in my room, Ghodrat (*his friend*) always sat here.

The director also tries to add annoying noise of cars and their lights to ferment the audience's hatred of these constructions and the demolition of the old house.

Fig. 5.2 A man is standing in the middle of an urban highway, where his old house was—*Soltan* (Kimiaei 1996. Reproduced by permission of Masoud Kimiaei)

The vertical growth of Tehran, as shown in Soltan, and the horizontal growth, as shown in *The Tenants*, are established as important parts of the visual memory of the audience about the city. Perhaps few people of today's generation remember Tehran's beautiful foothills and it is the cinema that can offer a clear picture about the breaking down of the last frontiers of Tehran. The image of middle class residents of the city outskirts which started with *The Tenants*, continues with the image of poverty and other social problems of lower class residents in many films such as *The Blue-Veiled* (Bani-Etemad 1995) and *The Beautiful City* (Farhadi 2004). These movies showed a 'City of the Marginalised'.

The first shot of The Blue-Veiled is one of the best images of these marginal neighbourhoods: a view from furnaces while Tehran is visible in the background. Such targeted shots can be seen throughout the film: sleeping mats and minimum living facilities belonging to jobseekers in a poor area. To specify boundaries of a marginal zone by rail tracks and trains is common. It seems that passing trains with their annoying noise in every moment remind marginal residents that the city still has not accepted them.

Sometimes directors prefer to show urban slum dwellers instead of marginal areas. In *The Girl in the Sneakers* (Sadr Ameli 2001) slum dwellers are the last refuge for a girl who has escaped home. This film reflects a city incapable of harbouring a single girl. The girl's truancy is flaneuristic at first, but then she walks aimless. *Boutique* (Nematollah 2003) is also about a girl who roams in the city alone. Different spaces and elements in the city shape the physical context of the story; from north to south of the city, with Milad-e-Nour shopping centre, Hassan Abad square, the red tower in Argentina Square and metro stations among them. Both *Boutique* and *The Girl in the Sneakers* condemn such truancy in city and thereby the city itself. One of the final shots of *Boutique* simply shows the subway train's tremendous momentum. This shot, along with the final shot showing a tower's lift in movement represent the speed and machine life of the soulless city.

Sometimes, this 'mirage city' changes to 'panic city': the city not only unable to harbour runaway girls, but also threats and kills citizens who work, study and live in it. *The Nights of Tehran* (Farhang 2001) is one of these films. Poverty, vulnerability and psychological problems of modern life create a panic city for its inhabitants: a city with mysterious people and spaces most residents are unaware of. Tehran is a forest of wild men in this film, and the place hosting crimes days and nights, and under sunshine and rain.

Drug addiction is one of the main issues in the city that has become the theme of many films that show a city-image which can be called the 'City of Addiction'. Strangely, addiction is most common in middle-class and upper town residents. *Santoori* (Mehrjui 2007) represents such city. The film shows the drug addiction of a famous musician, called Ali. One scene shows Ali seeking drugs among addicts. This sequence finally cuts to a long shot of the city; a shot which shows Ali looking at a city that stripped him of everything (Fig. 5.3). The radio in Ali's hand reads daily traffic news, talks about a city that has forgotten its artist and made him an anti-hero. It seems that Ali speaks to the city in whisper. Even after he treats his addiction, he does not want to come back to the city:

Fig. 5.3 An addicted artist looks at the city—*Santoori* (Mehrjui 2007. Reproduced by permission of Dariush Mehrjui)

> Doctor! Can you do something for me to stay here? Here I can do something. For example, I can teach guys music. Please don't let me come back to that wild city.

But movies are not only about the city's middle and higher classes. Class conflict is another theme of urban films which show a 'City of Class Contradictions'. *Tehran, Tehran* (Mehrjui and Karampour 2009) is one of them. Two film directors, one old and the other young, present a socio-physical image of Tehran according to their views. In the first episode named 'Tehran: Acquaintance Days' love waves among people in the city. However, the film begins with a bad event and the audience see a family whose house is collapsed on the new-year day, but they meet lovely people. It seems that the director wants to introduce this loving city in the worst day of a family. He takes a homeless family to the city to enjoy its beauty.

The second episode named 'Tehran: The Last String' is directly in contrast with the first half of the film. It is about young men who enjoy rock music and rally racing. Unlike the first episode, there is no image from the old neighbourhoods and old palaces of Tehran. The 54-storey tower, Cinema Mellat, the Artists' House, and the Milad Tower are elements of Tehran in the second episode. But it seems that the city has disappointed young men and they are not in love with their city:

> Was not the amusement park at the end of this highway? But there is nothing now. Amusement parks are destroyed. Everything is destroyed in this city. Tehran is like the amusement park. You should go to the end of it as fast as you can.

But it seems that movies which have tried to infiltrate into Southern life and deal with their social, cultural, economic and physical aspects of life have been more successful in presentation of a realistic image from the southern areas: picturing an image we can call the, 'City of Southern Residents'. It seems *Café Setare* (Moghadam 2006) is one of the best Iranian film in this area.

The movie nicely tries to link the spirit of life in southern areas of the city with urban physical features. Like works in Film Noir style, a cafe is the location of meeting and encounters of the story. Café, car repair shop, shrine, old houses and the neighbourhood's people make the space of the story. Locals know each other and are very close. Poverty and unemployment are the most important problems, and can disrupt the normal life of the neighbourhood. Different aspects of the arrival of wealth and modernity also threaten the neighbourhood.

It seems that the 'City of Southern Residents' is the city of problems; but *Mum's Guest* (Mehrjui 2004) shows a hopeful image from these areas. This film nicely shows semi-private spheres of Southern houses along with their residents, a space which is hard to find in better-off areas of the city. The apartment life and communications between neighbours in better-off areas of the city, which is in contrast with the semi-private sphere of *Mum's Guest* can be seen in *Tambourine* (Bakhtavar 2008). The resistance of a resident of the apartment against satellite antenna installation, dispute between residents of the complex, disrespect for ethics and laws of the apartment, conflicts between public and private interests, etc. are some of the problems facing neighbours. Entering the apartment complex makes the audience familiar with diverse and conflicting views of residents symbolising middle classes of society, who are not what they seem.

Some directors present accidental encounters and alienation by spreading the space of story from an apartment to the whole metropolis. In fact, they pick characters as pawns in such a way that they encounter each other in the most critical moments without knowing each other. In other words, it can be said that in these films there are some order in the chaos of the city which determines the fate of urban residents in what can be called the 'City of Anonymous Encounters'.

Intersection (Davoudi 2006) is the story of urban residents who affect each other's lives without knowing it: People with modern appearance and activities experiencing urban life of the new century. Foreign travel, working as a gynaecologist, recording TV programs, parties and nightlife are some of these activities shown in shots of a restaurant, a university classroom, an airport, a modern shopping centre, etc. As usual, extremely long shots from the city and its structures help director make physical context of the story. But the turning point of the story happens in an urban highway where fates of its characters join together: people who would have walked past each other anonymously in the following morning, but an accident affect their whole lives. It seems that the 'intersection' of people and their lives is an undeniable feature of the metropolis.

5.7 Conclusion

Cinema as a global media has an important role in socially and culturally illustrating communities. City as the most important social phenomenon has been the main context for most films since the invention of cinema. Therefore, images of cities in cinema can be a unique source for urban researchers, making it possible to have an imaginary travel to many cities in the world.

Tehran is no exception, and it has had a major role as the location of films in the Iranian cinema. This chapter attempted to review some of the images of Tehran in Iranian cinema. As mentioned before, all images of Tehran in cinema have focused on its problems. It means that Tehran always has been seen as the location of suffering, hardship and concerns. City of immigrants, city of refugees, city of horizontal expansion, city of tall towers and speculation, city of runaway girls, city of addiction, etc. were some of the images of Tehran in cinema. Although there are parallels to this in other nations' cinema, it seems that these very bleak pictures in Iranian cinema are in need for further analysis and research.

The only film that presents a utopian image of Tehran is an animation called *Tehran 1500* (Azimi 2011). In it there are multi-storey highways, cars that fly rather than move on the ground, human-like robots that live in the city, a multi-storey cemetery, travelling to other planets to continue education, and towers occasionally much taller than the Milad Tower. Keeping some traditional customs, religious behaviours, and the Azan voice are elements and activities that make Tehran in future. It seems that Tehran needs such images and perception about its future.

References

A night in Tehran (2008) Film. Directed by Bahram Kazemi [Original Title: Shabi dar Tehran]
Beautiful city (2004) Film. Directed by Asghar Farhadi [Original Title: Shahr-e-Ziba]
Boutique (2003) Film. Directed by Hamid Nematollah
Café Setare (2006) Film. Directed by Saman Moghadam
Canary yellow (1988) Film. Directed by Rakhshan Bani-Etemad [Original Title: Zard-e-Ghanari]
Croot P (1997) Cinema and architecture: melies, mallet-stevens, multimedia, by Francois Penz and Maureen; translated by Jafarinejad, Shahram (2009); Soroush press, Tehran
Fakouhi N (2004) Urban anthropology. Nashreney press, Tehran
Hutchison R (2010) Encyclopaedia of urban studies. SAGE publications, London
Intersection (2006) Film. Directed by Abolhasan Davoudi [Original Title: Taghato]
Leaning on the wind (2000) Film. Directed by Dariush Farhang [Original Title: Tekye bar Bad]
McQuail D (2000). Mass communication Theory, translate by Ejlali, Parviz (2009), Ministry of culture press, Tehran, Iran
Mennel B (2008) Cities and cinema. Routledge, New York
Mirage (1986) Film. Directed by Hamid Tamjidi [Original Title: Sarab]
Mum's guest (2004) Film. Directed by Dariush Mehrjui [Original Title: Mihman-e-Maman]
Santoori (2007) Film. Directed by Dariush Mehrjui
Shiel M, Fitzmaurice T (2001) Cinema and the city. Blackwell, Oxford
Soltan (1996) Film. Directed by Masoud Kimiaei
Tambourine (2008) Film. Directed by Parisa Bakhtavar [Original Title: Dayereh Zangi]
Tehran 1500 (2011) Film. Directed by Bahram Azimi
Tehran, Tehran (2009) Film. Directed by Dariush Mehrjuy and Mahdi Karampour
The blue-veiled (1995) Film. Directed by Rakhshan Bani-Etemad [Original Title: Rousari Abi]
The girl in the sneakers (2001) Film. Directed by Rasoul Sadrameli [Original Title: Dokhtari ba Kafsh haye Katani]
The nights of Tehran (2001) Film. Directed by Dariush Farhang [Original Title: Shab haye Tehran]
The tenants (1986) Film. Directed by Dariush Mehrjui [Original Title: Ejareneshinha]
Vasle-e-nikan (1991) Film. Directed by Ebrahim Hatamikia

Part II
Contemporary Society and Culture

Chapter 6
From Modesty to Immodesty: A Quranic Reading of Change in the Islamic Iranian City

Seyed Mahdi Khatami and Michael Tawa

Abstract To comprehend traditional Islamic urbanism, and draw lessons for contemporary practice, a deep conceptual understanding of the philosophical and religious parameters that influenced the development of Islamic cities is essential. Nowadays, traditional Islamic urbanism is appreciated and promoted in academic circles and architectural schools, while in many cases its lessons are ignored in practice. A mitigating factor is that traditional urbanism is considered merely in terms of the physical aspects of the city, whereas a city in fact contains and makes human lives. These two dimensions of the city reciprocally affect and inform each other and are both necessary to give a better understanding of Islamic urbanism, its historical traditions and its future. Urban paradigms can be studied and understood according to their principles and physical structures under two categories. One relates to specific historic periods and their cultural norms and associated technical and ecological conditions. The other is timeless and invariable across various eras. One of the significant timeless concepts, with its roots in Islamic philosophy and value systems, differentiating the character of today's cities from traditional settlements, is *modesty* in traditional urban fabrics. This essay aims to identify the intellectual basis of this theme in Islamic beliefs, to investigate its roots in *the Quran* as the core of the Islamic tradition, and to venture some of its possible urban and architectural implications. The essay will suggest urban design principles and implications for those wishing to maintain traditional Islamic philosophical, religious and ethical precepts in contemporary urban design.

Keywords Islamic city · Modest city · Islamic urbanism · Quranic city · Iranian cities

---

S.M. Khatami (✉) · M. Tawa
Faculty of Architecture, Design and Planning, The University of Sydney, Sydney, Australia
e-mail: seyed.mahdi.khatami@gmail.com

M. Tawa
e-mail: michael.tawa@sydney.edu.au

F.F. Arefian and S.H.I. Moeini (eds.), *Urban Change in Iran*,
The Urban Book Series, DOI 10.1007/978-3-319-26115-7_6

6.1 Introduction

Classic conceptualisations and readings of the Islamic city are very much a Western affair. Such readings were initiated by French orientalists and pursued by the British, German, and later, American scholars in different parts of the Middle East (Al-Sayyad and Türeli 2009: 600). Besides the merits of these studies and their contributions, the majority of these scholars focused on physical aspects of Islamic cities and neglected the value system behind them. Despite this lack of deep comprehension of Islamic cities, the trend was influential enough to persist. A majority of architectural schools and designers accepted the orientalists' definition of the Islamic city in theory, even though it was different from the deep nature and structure of these cities. In 'The Islamic City: Historic Myth, Islamic Essence, and Contemporary Relevance', Abu-Lughod (1987: 160), a critic of this trend, stated:

> My own book on Cairo fell into the trap set by the orientalists by accepting many of the earlier authorities about the nature of the Islamic city. The edifice they had built over the years seemed to me a strong and substantial one. Only gradually did it become clear how much a conspiracy of copying and glossing had yielded this optical illusion. But criticisms were already beginning to appear.

Several decades after the emergence of this approach to the Islamic city, it became clear that analysing the experience of traditional processes was more crucial than direct analysis of physical organisation and built form. In this regard Hakim (1991: 274), emphasised the notion that 'we should not copy from tradition, but develop principles and lessons from a clear understanding of concepts, models and theories from man-environment studies as applied to traditional settlements'. In other words, instead of merely focusing on the physical characteristics of urbanism or studying the city as an artefact or product, the processes that formed and informed the city should be investigated and noted.

The process of Islamic city formation has been influenced by numerous forces such as technological capability, political system, terrain and climate, social cultures and organisations, Islamic beliefs and laws and so forth. Some of these forces are related to a specific era. For instance, technological means such as heating and cooling methods obviously are not timeless, because the built environment changes to correspond with new technologies. Therefore it is impossible to extend a form, which has been made by mainly a particular technical force in a particular time, to another era. In a similar way, some political and governmental forces affecting the formation of cities relate to specific periods, because with changes in political power, some of these forces and related attitudes have changed entirely. Therefore, the imitation of one urban pattern is unlikely to be relevant or useful to a different era. Furthermore, each place and region has particular characteristics related to its climate, terrain and specific culture and norms. Some classic scholars have repeatedly ignored such key issues, studied a few Islamic cities and taken their generalised findings as universal.

However, there are also some influences that do not depend on specific places or eras. These concepts can apply in diverse Islamic regions and during different periods of time. One of those significant immutable concepts with roots in Islamic philosophy and value systems distinguishing the character of today's cities from traditional settlements is the prevalence of the concept of modesty in traditional urban fabrics. Many scholars such as Al-Sayyad and Tureli (2009: 601), Mortada (2008: 69), ʿAlīābādī(2009: 64) and Naghizade (2006: 120) have noted modesty as a key theme of the Islamic city. This essay aims to identify the intellectual basis of this urban characteristic in Islamic beliefs and to investigate its roots in the Quran as the core of the Islamic tradition. A comparison of how the concept of modesty is reflected in traditional urban fabrics and contemporary cities of Iran will be shown to imply that a degree of immodesty is prevalent in the contemporary city. The essay will further introduce some urban principles, based on the theme of modesty, which can still hold good for contemporary urban design practice.

6.2 The Concepts of Modesty and Immodesty in the Quran

Four verses in three chapters of the Quran describe and suggest the concept of modesty by commenting on features of human *walking* upon the earth in general. Surah al-Isrāʾ (Chapter 17, Verse 37) and Surah al-Furqān (Chapter 25, Verse 63) convey a general command from the God about the modest way of walking, while Surah Luqmān (Chapter 31, Verse 18, 19) convey the advice of the sage Luqman to his son, counselling against immodesty in walking. A comparative analysis of these verses will help to understand the meaning of modesty in the Quran.

Chapter 25 (Surah al-Furqān), Verse 63 states: '*The (faithful) slaves of the Beneficent are they who walk upon the earth modestly, and when the foolish ones address them answer: Peace*' (Pickthall 2001: 365). In this chapter, twelve attributes of faithful individuals are listed, with modesty being the first and foremost among them. The word *Hawn* (modesty), is an infinitive, meaning softened, peaceful and without arrogance. Using an infinitive in place of posture adverb in the Arabic language serves as an emphatic. Hence, by modesty, the faithful walk in a way that directly manifests peace and lack of arrogance (Makārim Shīrāzī 1995: 147). To interpret this verse and define *Hawn*, Imām Sādiq, the sixth Imām of the Shi'a faith, stated: 'the meaning of verse refers to a person who walks on the basis of his nature and prevent from arrogance and pride' (Ṭabarsī 1993: 222). It is important to note that the concept of 'slaves of the beneficent' in this verse and the related concept of modesty are founded on the idea of worshipping and serving of God. They denote the proper disposition and countenance of worshipers, rather than their servility, weakness and abjection. These two concepts have an essential difference: one of them comes with peace and honour and the other with baseness and insult. In a prayer to God, Imām ʿAlī, the custodian of the knowledge of the prophet, says: 'Oh God, this honour is enough for me that you are my Lord, and this pride is enough for me that I am your slave' (Subḥānī 2001: 248). The verse implies the need to prevent arrogance and egoism that can appear in all human actions and manners, beginning with the mode of walking which is a preeminent sign of human ethical demeanour (Makārim Shīrāzī 1995: 147).

Chapter 31 (Surah Luqmān), Verse 18 and 19 states: 'And turn not your face away from men with pride, nor walk in insolence through the earth. Verily, Allah likes not each arrogant boaster. And be moderate (or show no insolence) in your walking (pace)' (Khan et al. 1999: 412). Luqman thus forbids his son from improperly walking and behaving towards people, advising him to be moderate in his walking. The terms used in this verse are significant. For example, *Tas'eer* (to turn [the face] with pride), is a word referring to a kind of sickness in camels which causes their necks to become crooked. *Maraḥ* (insolence) also means great pride, arrogance and self-satisfaction obtained due to wealth and position. *Mukhtāl* (arrogance) normally refers to persons who assume themselves superior on the basis of illusory or imagined grounds and *Fakhūr* (boasting) is normally used for a person who demonstrates pride to others. *Mukhtāl* refers thoughts and *Fakhūr* to actions (Makārim Shīrāzī 1995: 54).

Besides forbidding walking arrogantly and proudly due to over indulgence in wealth and position, Surah Luqmān implicitly associates this characteristic to a physical, external sickness and its mental, internal correlate (verse 18)—but both Tas'eer and Mukhtāl evidence themselves in the actions and quality of walking. The following ḥadīth illustrates the sickness in the words of the Prophet Muhammad:

> One day, while walking, the Prophet saw a group of people gathered at one point. He asked them about the reason why they were gathered. They answered: there is a mad person here and his crazy and funny actions have absorbed people's attention. The Prophet told them: Do you want me to introduce you to a real mad? Men were silent and were listening carefully. He told them: a person who walks with arrogance and pride and constantly looks on both sides of himself, moving his flank and shoulder and does not see anyone and anything except himself; a person that people do not expect him any benefit and are not safe from his evil; he is real mad; but the person here is just ill' (Makārim Shīrāzī 1995: 58).

Finally, Verse 37 of Chapter 17 (Surah al-ʾIsrāʾ) states: 'And walk not on the earth with conceit and arrogance (in insolence). Verily, you can neither rend nor penetrate the earth, nor can you attain a stature like the mountains in height' (Khan et al. 1999: 285). The first part of this verse forbids people to demonstrate arrogance by exulting their wealth and position through their way of walking. Wisdom arises where human beings understand their scale in relation to God; arrogance and pride being nothing other than an illusion which causes humans to struggle and compete with God, world and nature.

6.3 The Roots of Modesty and Immodesty in the Quran

The Quran implies that the recognition and acceptance by human beings of their standing in relation to God would discourage them from acting arrogantly and boastfully while encourage them to be humble. On the other hand misunderstanding of this fact causes the desire to be superior among others. Therefore, besides introducing modesty as a feature of faithfulness, The Quran lists exaltedness as an equal prohibition. In verse 83 of Surah al-Qaṣaṣ (Chapter 28) this attribute is mentioned as a direct cause of human downfall: 'That home of the Hereafter We assign to those who do not desire exaltedness upon the earth or corruption. And the [best] outcome is for the righteous' (International 1997: 395). This verse follows the narration of the story of Qārūn (Korah), which points out that the very rich high-handedness in dealings related to property and power. Therefore those that oppress and corrupt are liable to suffer God's wrath.

The Islamic philosopher and interpreter of the Quran Seyyed Muḥammad Ḥussein Ṭabāṭabā'ī (1995: 120) believes that exaltedness (*ʿuluw*) is a kind of corruption (*Fisād*). '*ʿUluw*' is the infinitive of '*ʿAlā*' as a simple past. '*ʿAlā*' means a thing or person who was or became high, elevated or lofty (Lane 1968: 2142). This exaltedness can refer to wealth and position or to the possession of a huge palace or very small belongings. A number of interpreters of the Quran have stated the following ḥadīth to define the meaning of this verse: 'Imām ʿAlī was walking in the bazaar and was opening the Quran and was reading this verse and was saying: "This verse has descended for oppressor rulers and also modest powerful men" (Ṭabarsī 1993: 243). In this quotation, exaltedness in leadership is not rejected outright. Powerful rulers such as Imām ʿAlī did not accept power for exaltedness and therefore were extremely modest during their leadership. A leader who realises their position never tries to be boastful and proud because they understand the greatness of the God. This understanding accompanies believers in all their doings and prevents them from becoming egocentric since they will necessarily always remember God in different times and situations. In this regard Chapter 3 (SurahĀli ʿImrān), verse 191 of the Quran states: '…remember Allah, standing, sitting, and reclining, and consider the creation of the heavens and the earth, (and say) Our Lord! Thou create not this in vain. Glory be to Thee! Preserve us from the doom of Fire' (Pickthall 2001: 75).

Satan is a case in point. After 6000 years of worshipping God, Satan's selfishness and boastfulness led to exaltedness, expulsion from paradise and creaturely misery. When God ordered him to prostrate to Adam, he answered: 'I am better than him, you created me from fire, and you created him from clay.' And then God said: 'Then get out of here, for verily you are outcast. And verily! My Curse is on you till the Day of Recompense' (Khan et al. 1999: 152). In this regard Imām ʿAlī, in Nahjul-Balaghah in his sermon of 'Qāṣiʿi' about the selfishness and pride of Satan stated:

> His vanity stood in his way. Consequently, he felt proud over Adam by virtue of his creation and boasted over him on account of his origin. Thus, this enemy of Allah is the leader of those who boast, and the fore-runner of the vain. It is he who laid the foundation

> of factionalism, quarrelled with Allah about the robe of greatness, put on the dress of haughtiness and took off the covering of humility. Do you not see how Allah made him low on account of his vanity and humiliated him for his feigning to be high? He discarded him in this world and provided for him burning fire in the next world (Nahjul-Balaghah 2009: 624).

In fact pride causes illusion and refutation of reality. It is fundamentally against the nature of creatures like human beings. On this point, Imām Sādiq stated: 'a man's decrease in wisdom is proportional to the extent of pride in his heart.' In addition, according to the Prophet, 'a person who has a little bit of pride and arrogance does not enter into paradise.' The reason that this person has been deprived from paradise is that the consequence of pride is atheism, denial of the truth and of the sanctity of the divine. A person asked Imām Sādiq: 'what is the first step of denial of fact and atheism?' He answered: 'arrogance and pride' (Subḥānī 2001: 244). Therefore Quran states that 'the paradise is assigned to those who do not desire exaltedness upon the earth or corruption' (International 1997: 295).

6.4 Extension of the Concept of Modesty to Urban Fabric

While verses cited above concerned the concept of modesty in walking as a demonstration of a manner of being upon the earth, this concept is extendable to urban fabric. The Quran interpreter Nāsir Makārim Shīrāzī (1995: 147) states that such verses show the obscenity of insolence and arrogance even in a simple action like walking upon the earth. When verse 18 of Surah Luqmān states 'Verily, Allah likes not each arrogant boaster' (Khan et al. 1999: 412), it forbids both an inward manner of arrogance and its outside correlate, exemplified in the rhythms and modes of walking, the kinds of clothes worn and the methods of building. In Chapter 17, Verse 37 the matter of appropriate scale that respects nature and other created beings without arrogance is foregrounded.

Moreover in some ḥadīths and verses of the Quran, instances related to architecture and urban characteristics are mentioned. Verse 83 of Surah Al-Qaṣaṣ, which is about exaltedness upon the earth, identifies the wealthy citizen Qārūn, demonstrating his exaltedness through the palace he builds, and being punished for his immodesty: 'So We caused the earth to swallow him and his dwelling-place. Then he had no host to help him against Allah, nor was he of those who can save themselves' (Pickthall 2001: 395).

The next verse confirms that immodesty in relation to house and property was the reason for his downfall:

> And those who had wished for his position the previous day began to say: Oh, how Allah extends provision to whom He wills of His servants and restricts it! If not that Allah had conferred favour on us, He would have caused it to swallow us. Oh, how the disbelievers do not succeed! (International 1997: 395).

In the interpretation of verse 83, Imām Sādiq states 'the intention of 'uluw, which has been translated to exaltedness, is superior building'; and the Prophet Muhammad says:

> A person who constructs a house for showing-off and for fame, God in the hereafter puts it on his back whilst this house is a conflagrant fire. People asked him: 'Oh prophet! How he constructs a house for show off and fame. He stated: 'He built it over his need and adequacy to show his arrogance upon his neighbours ('Āmilī 2009: 62).

6.5 Main Features of Modesty in the Islamic City

Reading across an investigation into traditional, premodern Islamic cities, verses from the Quran and the ḥadīth, four key design attributes that convey notions of modesty and humility emerge: simplicity, unity/harmony, human scale and introversion.

Simplicity: In several ḥadīths, the Prophet orders his followers to dress simply and to avoid drinking and eating in gold and silver containers. Likewise, the Prophet specifies a simple building for the first mosque at Medina, without decoration or ornamentation; and when his followers suggest decorating it he does not accept. From this point of view, the urban fabric is the simple and un-luxurious cloth of the city that by implication dresses the mosques and all its buildings equally. However beautiful and spiritually elevating, the mosque must remain simple and without apparent luxury (Nubahār 1997: 81). Islamic value systems emphasise beauty as well as simplicity: 'The God is beautiful and likes beauty and liked to see the influence of this blessing on the people' (Nadīmī 2007, p. 105). Hence, and according to Titus Burkhardt, simplicity does not prevent beauty, nor does it result in monotony (Burckhardt 1993: 84).

Unity and Harmony: In as much as modesty conveys the correct relationship between the creator and the created, in the Islamic city, none of the individual buildings that form the urban texture should exalt itself or take prominence since all the parts should make up a whole in relation to the divine. This approach produces unity over and above changing circumstances of time and place. No building seeks to overcome or be higher than others so that a unity of form and skyline is achieved. Likewise, no building expresses distinctive materiality so as not to 'boast' in separating itself from others, again in order to emphasise a resultant unity.

Human Scale: Human scale in urban fabric is one of the key qualities stemming from modesty. Muḥammad Karīm Pīrnīā, master of traditional architectural scholarship in Iran, considers *mardumvārī* to be a primary principle of Iranian architecture. In his point of view Mardumvārī means: 'considering proportions between building organs and human organs and paying attention to human needs in building works' (Pīrnīā 2008: 26). Therefore, in the Islamic city everything is built in proportion to human dimensions in order to convey respect and modesty in relation to the God's Creations. Moreover, some ḥadīths forbid constructing a building more than about four meters (eight zirā') in height. Even for a mosque, which is a pivotal component of the Islamic city, it is considered detestable (*makrūh*) to build walls higher than needed and thereby displaying pride and boasting ('Āmilī 2009: 74). In one ḥadīth, the Prophet states: 'I have not been assigned to build mosques high.' Another, from Imām Ṣādiq, explains that the height of the Prophet mosque in Medina was the same height as a person (Nubahār 1997: 89).

Introversion: A fourth attribute of the Islamic city that stems from both modesty and privacy is introversion, which some scholars consider as a defining feature (Pīrnīā 2008: 35). This trend can be seen in mosques whose outside walls have less decorations than their interior spaces, thereby focussing predominantly on interiority as the major characteristic of worth (Naghizadeh 2006). Such introversion applies as much to residential buildings around a courtyard as it does to the introversive orientation of the mosque. Such orientations respond to multiple registers: functional, ethical, authentic and symbolic, with implications for privacy, modesty and beauty.

6.6 Change in Islamic Urban Fabrics: From Modesty to Immodesty

After colonisation of Islamic countries by Western countries, foreign cultures and ideologies had a significant impact on the Islamic context. This humanist worldview expanded into different areas, some political rulers in Iran imposed Western paradigms and their associated aesthetics with a view to championing Western-style development. At the same time, several orientalists started to theorise the fast disappearing of the Islamic city by introducing simplistic or incorrect assumptions, claiming them to be original to the traditional Islamic city. In some cases, dynastic styles of architecture were introduced as a way of recovering lost traditions in spite of the fact that traditions are not defined by building styles and that the reality of the traditional Islamic city is not wholly contained in its form. Further like contributions were made by Iranian graduates of Western architectural schools who sought to apply external paradigms in the guise of either recovering a lost tradition or conveying the country into a nascent democratic capitalism through architectural types solely developed out of Western concerns and contexts.

A corollary of these attitudes is that the concrete world is considered as an ultimate goal, and that different cultural aspects—such as those attaching to ethical, religious, sacred or symbolic matters for example—are devalued. In this approach, designing becomes a struggle for the exaltedness of architect and designers who deploy their skills to demonstrate power and capability or to confirm their status among designers and architects.

From a Quranic point of view, as mentioned in Chapter 17, Verse 37, the cause of these tendencies is the incorrect imagination of the scale of human beings and thus a lack of modesty. Therefore to clothe oneself in such a way as to draw attention to oneself or to separate oneself from others is forbidden in Islamic jurisprudence. This cloth has its roots in arrogance, pride and glaring. In a same way and for a same reason, in the Islamic point of view, the cloth of a city should not be for glaring, but all buildings should cover it with a colourless cloth in form, shape, size and decoration (ʻAlīābādī 2009: 95).

The lack of modesty, through this struggle for arrogance and glaring, causes immodesty in front of God, nature, people and the city. Its main consequences for the urban fabric features are:

- Lack of unity and harmony with each part equally seeking to dominate
- Trend to luxury and consumerism associated with a lack of simplicity
- An extraverted attitude to design which overturn the possibility of introversion as a reflection of modesty
- Inattention to nature and natural order in favour of overtly human and technological concerns.

6.7 Conclusion

To draw urban lessons from traditional Islamic urbanism for contemporary practice, a deep understanding of concepts, values and beliefs behind the formation processes of these cities is necessary. One of the significant concepts distinguishing the character of today's cities from traditional settlements is the prevalence of modesty and humility in traditional urban fabric. This concept has a strong root in the Islamic value system, particularly in the Quran as the most important religious book of all Muslims.

The four main attributes of traditional, pre modern Islamic cities which respond to the concept of modesty and its associated qualities have been suggested as simplicity, harmony, human scale and introversion. The Western cultural invasion represented by modernity brought with it new attitudes that emphasise individuality over community and a focus on the concrete world as the sole reality. These have influenced the conceptualisation and practice of urban design in Iran, bringing with them an arrogance and extroversion in contrast with the modesty that was once valued and codified in human behaviour and the traditional urban fabric. The consequent immodesty of contemporary cities can be characterised by a lack of unity and harmony, a tendency for the individual expression of isolated buildings, a related tendency towards luxury and consumerism, and outward concerns that overlook the inner, deep order and structure of nature and the created world.

Several verses in the Quran describe and suggest the concept of modesty and humility and its demonstration in human behaviour such as walking. The roots of this modesty are related to submitting to God and understanding the role and the position of human beings in Islamic faith. Understanding the scale of human beings compared with God's power and His creations strongly encourages people to be humble and modest and mitigate arrogance and boastfulness in relation to others. Incomprehension of this reality causes the desire for superiority over others. Therefore besides describing modesty as a feature of faithful people, the Quran forbids exaltedness and haughtiness that would isolate individuals from the community of the faithful. These moral conditions and exhortations have possible extensions into urban fabrics and are not limited to human behaviour, since the style or manner of human beings has a correlate in the style or manner of building.

References

Abu-Lughod JL (1987) The Islamic city—historic myth, Islamic essence, and contemporary. Int J Middle East Stud 19:155–176

ʿAlīābādī M (2009) Furūtanī yā Hunar-i Bandigī dar Miʿmārī-i Zamīn. Hamāyish-i Millī-i Pazhūhishhā-yi Qurʾānī-i Ḥuzi va Dānishgāh, pp 61–112

Alsayyad N, Tureli I (2009) Islamic urbanism. In: Rob K, Nigel T (eds) International encyclopedia of human geography. Elsevier, Oxford

ʿĀmilī JM (2009) Shahr-i Islāmī, Mash'had, Sāzmān-i Farhangī Tafrīhī-i Shahrdārī-i Mash'had

Bābāʾī AA (2003) Guzidih-i Tafsīr-i Nimūnih. Dār al-Kutub al-Islāmīah, Tehrān

Bemanian MR (2007) Some approaches toward explaining muslims architectures indicators. Town council organisation, Tehrān

Burckhardt T (1993) An introduction to principles and methods of religious art. Article collections of Titus Burckhardt, Tehrān

Burckhardt T (2010) Fez: city of Islam. Ḥikmat, Tehrān

Hakim BS (1991) Urban design in traditional Islamic culture: recycling its successes. Cities 8:274–277

International S (1997) The Qurān: Arabic text with corresponding english meanings, ideas4islam

Khan MM, Khan AH, Taqi-Ud-Din M (1999) Noble Quran. Dār al-Salām Publications, pp. 9960–740 (ISBN 978)

Lane EW (1968) An Arabic-English lexion

Makārim Shīrāzī N (1995) Tafsīr-i Nimūnih. Quran interpretation. Dār al-Kutub al-Islāmīah, Tehrān

Mortada H (2008) Traditional Islamic principles of built environment. Vizārat-i Maskan va Shahrsāzī, Tehrān

NadīmīH (2007) Kelk-e-Doost: Ten Esseys on Art and Architecture. Sāzmān-i Farhanī Tafriḥī-i Shahrdārī-i Iṣfahān, Iṣfahān

Naghizadeh M (1997) Principles of contemporary Islamic urban design. The elimination of conflicts in Muslim built environments based on order and justice: The experience of Semnan, Iran. Ph.D., University of New South Wales (Australia)

Naghizadeh M (2006) Islamic city and architecture. Mānī, Iṣfahān

Nubahār R (1997) Kūy-i Dūst, Tehrān, Daftar-i Muṭāliʿat-i Tārīkh va Maʿārif-i Islāmī

Pickthall MM (2001) The glorious Qur'an: the arabic text with a translation in english, TahrikeTarsile Quran

Pīrnīā MK (2008) Sabk Shināsī-i Miʿmārī-i Īrānī. Surūsh-i Dānish, Tehrān

Subhānī J (2001) Simā-yi Insān-i Kāmil dar Quran. Markaz-i Intishārāt-i Daftar-i Tablīghāt-i Islāmī, Qom

Ṭabarsī FEH (1993) Majmaʿ al-Bayān. Tehrān, Nāṣir Khusru

Ṭabāṭabāʾī SMH (1995) Tafsīr-i al-Mīzān. Qom

Chapter 7
Tehran: A Call for Spatial Justice

Zahra Azizi and Mahya Fatemi

Abstract Placed among a blend of geographic collages and augmented landscapes, societies are continuously striving for security, equality and drawing fair and democratic boundaries while injustice embeds itself into space. The phenomenon of the 'city' and 'urban life' is born through the evolving process of the satisfaction of social primary and secondary needs. Overcoming basic needs in life, calls for a profound personal spiritual fulfilment, sometimes seeking a powerful definition of existence. Needs lead to the formation of traditions that entail the repetition of desires. Through the continuity of a functioning social structure, urban fabrics come to being. Consequently, space is socially produced along with justice over time. Just or unjust behaviour manifested into space is a clear result of our decision making; whether personal, economic or political. Implementing one of the most precious yet neglected factors of human rights, the freedom to shape and reshape ourselves and the city, a collective force beyond that of the lone individual is required. Tehran, welcoming rural migrants for decades, is a collection of a widening social gap through various divisions including class and social background. Like every living entity a city can reach the brink of a nervous breakdown and there is a breaking point, for Tehran, that point is *now*. If she is pushed too far, one day she might just push back. For Tehran this chapter aims to extract opportunities, situations, ideals and dreams of a nation with an eye for beauty and a taste for socio-economic happiness through the lens of *seeking spatial justice*.

Keywords Right to the city · Tehran · Spatial justice · Social structures · Uneven development

Z. Azizi (✉)
Akaran Architects, University College London, London, UK
e-mail: zahra.azizi@gmail.com

M. Fatemi
The Bartlett Development Planning Unit, University College London, London, UK
e-mail: Mahya.fatemi@gmail.com

F.F. Arefian and S.H.I. Moeini (eds.), *Urban Change in Iran*,
The Urban Book Series, DOI 10.1007/978-3-319-26115-7_7

7.1 Introduction

'Space is not an empty void. It is always filled with politics, ideology and other forces shaping our lives, challenging us to engage in struggles over geography' (Soja 2010: 19). Seeking Spatial Justice has different scales as it can extend from environmental injustices like uneven geographical impacts of climate and global warming on one hand and to human body on the other. Nevertheless these are the two outer layers of extreme models. We intend to look at the 'in-between spaces' of these thresholds to geographical uneven developments. We will focus on urban examples of how and where (in)justices occur, to position our research for spatial justice in socially produced contexts.

Justice or injustice, and collective struggles for equitable social resources that a city can provide, has become an important matter of urban life, as most of the world's population resides in cities today. Urban life layers exist in various contexts below or above the administrative city space itself, which presents a multi-scalar view of the city with global, regional, national and localised scopes. This multi-scalar view of the city is the key to understanding the interpretive power and dimensions of a major spatial perspective, and essential in expansion of spatial theory of justice and injustice. Geographies of the unjust can be created by three different platforms of bottom-up or top-down social actions or in-between spaces where the local and the global meet. The goal is not to go in detailed topics but to illustrate a range of contexts that can be examined from a spatial justice perspective in Tehran.

Spatial justice in a city is closely linked to '*the right to the city*', associated with Henri Lefebvre. Lefebvre's dispute that '*the city is an oeuvre—a work in which all its citizens participate*' (Mitchell 2003: 17) very much defines the city as stage and inhabitants as actors almost going by an urban script. Diverse people with diverse agendas struggle with each other, to access the public realm and to equal the rights of citizenship. From this struggle 'the city as an ouvre, as a collective if not singular project—emerges and the new modes and models of living and inhabiting are invented' (Mitchell 2003: 18).

7.2 Setting Spatial Justice in Distributional Inequalities and Needs

Factors such as institutional inefficiency, racial prejudices, budget requirements, personal greed, differential power and wealth add to primitive distributional inequality, producing discriminatory geographies of accessibility to public health and services. Distributional inequalities can extend from critical public services such as mass transit, education, crime prevention, and police to more privatised supplies of suitable housing, employment and food.

These are spatial injustices that go against urban based civil rights and constitutional assurances of justice. Many actors have made these decisions to benefit mostly the powerful and the rich. In every city in the world, fully capitalist or not, the social geography of class persists to be spatially unjust. These inequalities are hidden under the affirmation that they are unavoidable, normal and expected consequences of urban life. This may have even been viewed committing to the greater public good and products of individualised freedom of choice (e.g. gated communities), an unquestioned formation of spatial advantage and privilege founded on differential power and wealth.

One of the first to uncover and examine this hidden urban geography of injustice is David Harvey (the book Social Justice and the City 1993). He argues that the network (decision making of banks, retailers, developers, real estate, labour, housing …) exertions of regulatory activities tend to constantly direct to redistribution of real income likable to the wealthy in an industrial capital city.

In order for us to elaborate, a reference to social needs becomes crucial. The city and urban life is born through the evolving process of the satisfaction of primary and secondary needs. Overcoming basic needs in life, calls for spiritual fulfilment or in other words secondary needs which on many occasions become primary for individuals who have moved beyond achieving rudimentary desires. The repetition of satisfying needs lead to the formation of traditions which in turn will be repeated; therefore the cycle goes on striving to maintain social structure. Within this cyclic process comes the essence of social justice which determines the way social institutions will work in relation to one another; whether or not they will work smoothly together and how their interactions and decision making will affect the formation of society.

The bourgeois city is not so much a place of participation but of expropriation by a set of economic interests and a leading class that is not interested in creating a city for cohabitation of differences. '*More and more the spaces of the modern city are being produced for us rather than by us.*' (Mitchell 2003: 18). To reverse this 'The need to creative activity (not only of consumable material goods), information, symbolism, the imaginary and play is significant' (Lefebvre 1996[1968]: 147).

More closely 'the right to the city' implies: *right to freedom, to individualisation in socialisation, and to inhabit. It also suggests the right to participation and appropriation.* (Lefebvre 1996[1968]: 174). The right to housing and inhabiting the city needs more than apartments and housing; it requires the betterment of the city in a way responsive to the demands and desires of inhabitants, particularly to the oppressed. Ayatollah Khomeini, the leader of the 1979 Islamic revolution stated:

> Value the nation. Serve especially the unprivileged who are strong pillars of society. The Islamic republic is brought to existence by their devotion and loyalty. Do not neglect them. Acknowledge them as part of one family; that of your own. (Translated from the booklet Tehran on the Path of Growth and Justice, Municipality of Tehran Press, May 2012).

And during its 35-year life span, this has been the frontier motto of the government towards the aim of providing just geographies.

Iris Marion Young (1990) argues that distributive justice is crucially necessary, but it is not adequate. Adding to a just distribution and content of social justice, a structure that allows complete, desired participation of oppressed groups for decision making and forward criticism of any forms of oppression must be included (1990: 48–63). According to Young (1990: 37), with regards to social justice concerns, two 'values' are necessary for a society to develop 'the good life':

- Developing and exercising one's capacities and expressing one's experience.
- Participating in determining one's action and the conditions of one's action.

Young realised that these are *universal* values and they need promotion for everyone; but there are values that *demand* cautious attention to difference as against them rest '*two social conditions which define injustice: oppression, the institutional constrain on self-development, and domination, the institutional constraint on self-determination* (1990: 37)'. Domination and oppression are both applied through difference: It is differently positioned actors who dominate and difference that is oppressed. The recognition of differences together with social promotion and autonomy is required to reach freedom. In the Iranian framework of life and in Capital Tehran, for many matters especially ideological and religious, it has been tough to maintain a sense of autonomy, of being truly oneself in public and in private. This seems to have caused an active sense of segregation, pessimism, and disrespect for differences in society.

For Young (1990) autonomy needs not just distribution of opportunities and goods, but improved socialised control over the *ways* of distribution. This control must be linked with normative, elaborate and universalising institutional framework, encouraging difference and autonomy of individuals and groups. Such formation of rights depends on Young's (1990: 98) notion of '*logic of identity*' that '*denies or represses difference*' because the logic of identity envisions entities in terms of substance instead of relations or process. Moreover '*logic of representation*' focuses on the right of individuals and groups to make their needs known, to represent themselves as legitimate applicants to public consideration even through struggles (Harvey 1996).

Space of representation, a location where individuals and groups can make themselves noticeable is vital. It is here that their desires can be noticed, opposed, or recognised. The logic of representation stresses the creation or betterment of social production of certain types of public space (Mitchell 2003: 33). Even though majority of states will closely control such spaces to maintain self security, still the functioning of these spaces helps state democracy which in turn ensures a great deal of justice. Of utmost importance is the provision and planning of public spaces. A public space of representation as such in Tehran is scarce (Fig. 7.1). 'What makes a space public—a space in which the cry and demand for the right to the city can be seen and heard—is when a pressing need is fulfilled. Representation both demands space and creates space' (Mitchell 2003: 35).

Fig. 7.1 A public space in Tehran (Authors 2012)

7.3 The Right to Services and Private Property

In market-based economies, every space has been commercialised into bundles of valued territory owned by state, individuals or corporations. Attention to property rights are needed to understand how unfair geographies are created. The right to the city and human rights are subordinated to the dominance of rights to property.

'There are zones of contention between public and private property rights and focal points for social action aimed at assuring residents' right to the city, in the sense of collective access to the common pool of public recourses the city provides' (Soja 2010: 46). Coalitions like environmental justice, community-based regionalism and building for community development need to be created on a regional or metropolitan scale. Access to public services and goods, is associated with regional rights to the city. Scales of spatial justice are not apparent as they interweave and interact in complex patterns.

The intention is not to formulate collective ownerships as a solution against private property ownership, but to introduce a fresh look with critical spatial perspective at the topic of private against public space and to achieve greater socio-spatial justice by exploring possibilities of new strategies. The purpose is to raise awareness of political organisation of space as it is forced from top down as a type of social control and kept by the land market, local state and the legal system.

Since the 1979 revolution and the fall of the monarchy, the main state objective became full economic independence, employment and better, more just living standards; which all seem reasonable steps. Land reform process which had already started under the Shah was to be continued free of political interests. The revolution brought forward the emergence of many self-driven organisations such as Construction Jihad (mobilisation for construction) solely for the fulfilment of public welfare. Almost an independent governing body for villages, it acted as a one-stop shop for provision of all sorts of services to rural areas from agriculture, electricity, water, roads, telecommunication and healthcare to family planning. Also, the Literacy Movement provided cumulative ascendant rates of literacy while also providing education for the deprived which served well for villages all over Iran. Although all these efforts could be classified under the promotion of social justice which in turn would gradually, at least hypothetically, be expected to manifest itself in the form of spatial justice, the situation changed over time. These organisations gradually dissolved and their responsibilities were overtaken by other governmental departments such as the Ministry of Agriculture and the Ministry of Education etc. Through diminishing and sometimes unsuccessful efforts of these social movements, or possibly through their lack of systematic coordination, many families migrated to Tehran from rural areas. Counterbalancing the Shah's selective methods for the distribution of power, the government of the Islamic Republic delegated many responsibilities to people from different social classes as a noble act.

Therefore, with the public's vast participation, albeit inexperienced, the ruling body ingeniously succeeded at least for the first 20 years after the revolution, especially in saving the country during the eight-year Iran Iraq war. With positive mottos and presumably good intentions, one would expect this process to have progressed over the years. In a metropolis such as Tehran, over a period of time when society lacks the appropriate institutions to overcome its needs, or rather these institutions are short of appropriate policies; a loss of trust will emerge which causes possessive individualism (Abbas Milani) to nurture itself amongst society. It also brings into being the formation of individual relations of a jobbery nature. Simple examples of this would be the purchase of certain products or services by starting close relationships with those holding these services, to make sure one can obtain and procure high quality merchandise etc. which means that acquaintances gradually become of a different nature; that of profiteering. Today, this has become one of the main factors preventing just actions and territories from emerging. One can almost see this as the advent of selfish acts and therefore the decline of selfless altruistic behaviour. An instance of this is the ever increasing effects of the '*Besaaz-Befroosh*' movement, where demand for carefully planned architecture is replaced by profit from construction regardless of aesthetics, regulations, and social welfare. This has become a problem embedded into Iranian spatial culture.

Entrepreneurial approaches in the planning process have created fierce competition among regions and cities for economic benefits, also fuelled by a decline in welfare oriented urban and regional planning. This is a contributing factor in the struggles for spatially manifested social justice.

7.4 'Privatopias' and the Role of Segregation in Spatial Justice

Another transformation that took place after the Islamic Revolution as a result of a profound change in values and ideologies was the change in authentic and functioning city fabrics. The state's discouraging attempts to eliminate centralised areas for the music industry, theatres and cinemas caused lanes such as the famous *Laleh Zaar* (close to Tehran's *Grand bazaar*) to go through decline and recession, in terms of appearance, economic gain through businesses and its urban function as a hub for the performing arts. During the Qajar Dynasty, *Laleh Zaar Street* was built through the royal gardens following Nasereddin Shah's travels to Europe and his desire for modernisation. It was intended to become the Champs-Élysées of Tehran with fashionable shops, cafes, theatres and à la mode women passing by. However with frontier businesses becoming derelict because of their non-compliance with the new revolutionary values, the street lent itself to electrical equipment shops. As wealthy families slowly moved out of the area it became an underprivileged neighbourhood in terms of quality of life, therefore creating an almost unjust territory; an area having lost its aura, class, identity left with a deteriorating architecture. Resistance to change has caused many urban areas as such to diminish; and the 'uptown' and the 'downtown' (in Tehran uptown or *baalaa-shahr* refers to the area of the upper class where downtown or *payin-shahr* is home to the working class) have almost swopped positions over the past four decades. This type of segregation among urban communities does not do the principles of spatial justice well as the gap widens and deepens over time if neglected; nor does it serve the purposes of cultural development of space which is one of the key ingredients of the establishment of social justice in urban space.

The role of the cultural understanding of urban spaces brings us to discuss *cultural diseases* which, as pointed out by Mohammad Beheshti, are nowadays often mistaken with culture (Ph.D. Students' Round Table with Darab Diba and Mohammd Beheshti, May 2012, Islamic Azaad University). Since 1963 when major migration from the rural areas to Tehran took place, well established neighbourhoods started to become polarised. By 1978 the existing social structure of Tehran was completely demolished. Tehran became a place defined by its population rather than social identity; in other words quantity rather than quality. This pleased urban planners as they had better comprehension of the quantitative rather than qualitative analysis. This also penetrated the urban vocabulary. Streets were now defined by their width, for example *8-Metri* or the *8-Meter Street* etc. It is as if people would be called by their birth certificate number, as if they never had a name.

7.5 City Walls: The Tangible and the Intangible

Various kinds of protective walls, physical along with psychological have emerged. The uneven effect of economic restructuring and globalisation has reached almost everywhere fortifying built environments. Residences have become even more guarded and covered in alarm systems, advanced security and surveillance, also many activities, everyday objects, land uses in the urban environment, libraries, shopping centres refuse bins and design spiked benches to hold off influx of the hungry and homeless. Are these gated communities intense expressions of democratic individualism and freedom of choice or are they spatially unjust? The answer is yes to both and is where the main problem occurs.

Determined by voluntary preferences and fear more people particularly from the upper income class abandoned the urban public life to reside in isolated 'Privatopias' (coined by Even Mackenzie 1994). This centrifugal trend away from the urban and city responsibilities is different to the so-called gentrification process in its absence of commitment to urban living. This privatisation is less explicitly dominated by the state which is another model of spatial colonisation.

The city has been drenched by surveillance cameras with the widespread obsessive sense of insecurity. Security obsessed urbanism has advanced especially in city regions where there is a larger population of poor and diverse ethno-cultural settlements of domestic inhabitants. The most assertive life-threatening expression of spatial injustice is this expanding gap between the rich and poor concentrations.

Today power, wealth and global distribution of resources are almost more polarised than ever before with the heightening number of superrich accumulated in a few favoured places whereas strikingly a billion people reside end-to-end in compact slums. These are virtually redlined areas of a city with drained income and savings which are perceived as unattractive places to live. Nowadays urbanisation, globalisation and its associated factors have spread unevenly.

'As the whole world becomes urbanised and globalised to some degree, the urbanisation of injustice and the globalisation of injustice reinforce one another to create what are probably the greatest spatial inequalities of wealth and power the world has ever seen' (Soja 2010: 60). Regionalised right to the city and seeking spatial justice is urgently needed. Theoretically the task of knowledgeable spatial practice is to bring to surface and public awareness the compositions of privilege, based on gender, ethnicity, class, disability, sexual preferences, race and any other structures of hierarchical control (Fig. 7.2). This is very much the case in Tehran. A city with evident concentrations of rich and poor, and highly scattered number of vendors than ever before which has been out of control in particular realms such as the metro; an area in need of an urgent social pathology and cure.

As time passes we witness a Tehran with walls going higher and more solid than ever, both in fear of crime and in making a statement for owners of buildings who use them as boastful objects. Nevertheless, positive steps have also been taken which mark the beginning of a long journey to fulfil relative social justice. The construction of the metro network has connected poorer parts of the city to the more

Fig. 7.2 'I am under control' (Authors 2012)

affluent where welfare facilities have been established a longer time which has made a vast variety of amenities more accessible. Also, the planning and fulfilment of more leisure, green, educational centres, discussed later, is promising. So long as the ultimate goal of any decision would be the persistence of social welfare, spatial justice would relatively be accomplished. However, the other end of the spectrum would be the immense gap between the affluent 'who seal themselves for protection' and the deprived who become poorer by default. This favours the wealthy and disfavours the unfortunate and therefore uneven developments form upon the geographies we live upon.

7.6 Time in the Eyes of (In)Justice

Taking the city as a living organism, one realises shortly after exploring Tehran, its gradual suffocation: '… moving towards a spatial theory of justice it becomes evident that socially produced geographies, because they are created by human actions can be transformed through human agency for better or worse …' (Soja 2010: 104). This suggests that the dialogue between Tehran and Tehraners has not always been constructive. By mishaps, weak decision-making and lack of rigour and knowledge, one can observe a disharmonic pattern of social and urban warps;

in both the physical and psychological sense of the word. It also seems that Tehran, responsive to its inhabitants, has lost track of keeping within a certain predefined size, of time and certainly neglected the significance of going by a carefully designed master plan.

Historically, many major cities used to define their boundaries by different means, mainly creating a ditch with running water in a ring around them. This was to keep the city immune. Over time one of the main ingredients of a healthy city in growth would be the possession of a green belt, which would protect the city while acting as its lungs. Over the past few decades of Tehran's urban development, the city's green belt has gradually faded giving room to a peripheral city which has been draining the energy out of the capital with intentional greed. Hence, we are now left with a Tehran of 14 million inhabitants; and instead of a green belt we have the city of Karaj (made capital of Alborz province in 2010) with a population of 7 million in its proximity. This further determines a hastening process of contaminated urbanisation for Tehran.

Since September 2005 when Mohammad Bagher Ghalibaf was elected Mayor, Tehran has seen itself greener than ever. As well as the city reaching its full capacity of highways for easier access and many other construction projects, many urban parks (*Boustans*) have come into existence. A *Boustan* is a large mass of the urban platform dedicated to greenery, social activities, public open and closed spaces such as amphitheatres, cafes, general sports, urban art, etc. These considerable plots of land have been planned mostly around unprivileged neighbourhoods to regenerate them, improving their quality of life. Some Boustans are even themed based on ancient Persian literature. An example of this is the newly inaugurated '*Javanmardan Boustan*' (Javanmard implies chivalry); it is a spatial courtesy to a selection of noblemen and women, those with courage, talent and wisdom (Javanmardaan Boustan Booklet for the opening night 2012). Figure 7.3 shows this. It is a reminder of the intrinsic need for justice and those who have actively worked towards it in the Iranian society. The hierarchy of the park revolves around these characters with help from parts of Ferdowsi's Shahnameh (The Book of Kings), a long epic poem based on the mythical and historical past of Iran written in the 11th century.

Not only is it pleasant to witness these urban spaces spread themselves over what many call a diseased city with an ill infrastructure, it comes to much delight to see the use of narrative in the designing of these parks rather than the conventional post-revolution Iranian approach of 'cut and paste culture' used by majority of protagonists in the construction industry. The care put into the planning, designing and execution of these parks mark the vivid persistence of social welfare as the goal of the officials and therefore the traces of the spatial manifestation of social justice. This process is the result of the early sparks of social and spatial consciousness which implies a new approach and the recognition of space as a volume filled with politics and privileges. Mayor Ghalibaf stated in regards to city justice: 'city management should deliver calm environments … to bring about peace of mind and comfort to people to enable them to advance in their journey of excellence' (Translated from the booklet Tehran on the Path of Growth and Justice 2012).

Fig. 7.3 Javanmardan Boustan (Authors 2012)

The Municipality of Tehran has recently declared their definition of justice under: 'Tehran on the Path of Growth and Justice' as follows:

> [...]Tehran is a city that has transformed during the last five to six years through macro and micro projects. Meanwhile, the great attention of the urban management to underprivileged parts of Tehran is a significant turning point. Today, the topic of north-south polarisation in Tehran is not being raised anymore. Projects such as 'Javadiye bridge', Towhid tunnel, cinemas and most importantly 'Boustan Velaayat Park' have placed Tehran on the route towards spatial justice. Moreover, the change in the approach to urban management by the municipality from a service provider to a social institution is one of the most just measures taken in the past few years (Translated from the booklet Tehran on the Path of Growth and Justice 2012).

One should give the Municipality credit for better roads, parks and green public spaces and efforts for lesser regional polarisation. Even though progress has been made, there still stands firm the notion of *Time*. Depending on situational factors, the city, social behaviour and justice are ever changing, revealing their temporal nature. This means that socio-political and economic decisions are also temporary and even though they may seem right in the heat of the moment, they may well be proven otherwise at a later stage. Many argue that the creation of parks and the regeneration of neglected urban spaces is all very well, but they do not undo the large-scale mistakes caused by past decisions. The 'Abbas Abaad Hills' for example had been planned as an ultra large-scale open green space, a valley which would collect Tehran's rainwater and flood streams, and a filtering lung for the city. However, these plans changed as officials ruled over to build mainly huge governmental and NGO buildings on the site as an act of greed. While nearly all the site was built up, a small park was planned on its corner, almost an act of shameless disguise. This along with many others is an example of not making the right choices at the right time. While it is never late for amendments it will take many times the energy to put things right.

Regarding city walls and murals with imagery of war martyrs (Fig. 7.4) and other political and religious (e.g. Quranic verses) subjects, we are presented with contrasting arguments. On one hand using urban walls for political propaganda and as a *space of (one-sided state) representation* to convey and enforce religious ideologies is somewhat unjust; but on the other hand the initial intention for these murals was for them to be a reminder of the unjust war imposed on Iran. The fact that most Tehraners do not notice these 'billboards' shows the government has been unsuccessful in delivering its message. The inclination of the state to impose its ideals through mechanisms such as billboards or intimidating patrols attempting to

Fig. 7.4 Imagery of war martyrs in Tehran Bazaar, (Authors 2012)

control public appearance is against freedom of autonomy which is a basic need for the prevailing of spatial urban justice.

> We make our geographies, for good or bad, just or unjust, in much the same way, we make our histories, under conditions not of our own choosing but in real world contexts already shaped by socio-spatial processes in the past, and the enveloping historically and socially constituted geographies of the present. This profoundly displaces the idea of space merely as container, a neutral stage for life's seemingly time driven drama (Soja 2010: 103).

The beauty of the passage of time, therefore, always remains surprising, fascinating, uncanny and sometimes unpredictable.

7.7 Conclusion

Hardly anyone cares to pick the mulberries any more. We speak from a generation who used to play hide and seek and run around vast gardens with spreads knotted between trees into which all the mulberries would fall in hot summer days. Playfully we would collect them, find a shaded spot and taste their sweetness with all the joy in the world. Today as we walk through the almost non-existent pedestrian sidewalks of Tehran, one can see branches of mulberry trees reaching out through the gaps between the carefully designed metal fences guarding *privatopias*, with berries falling and no one to attend or even care as they walk past and crush them under their heavy steps.

The solitude of mulberry trees is a minor example of the lack of sentiment, romanticism and poetics woven into the city fabric and hence our lives in Tehran. It is merely a symbolic representation of the insensitive transition from ruralism to urbanism. 'We all experience in one way or another the negative effects of unjust geographies. This makes struggles over space and the right to the city a potentially powerful source of shared identity, determination and effectiveness in changing the world for the better' (Soja 2010: 109).

> To confront social polarisation and exclusion, what is needed is to create possibilities for integration [...] The ability to create social integration, [...] should provide [...] access to material resources through access to work; to have access to decision making through political participation; and to have access to common narratives and identities through socialisation (Madanipour 1998: 255).

In an attempt to modernise and expand itself to provide space and primary needs for the bulk of dwellers and migrants flowing into the city in the past decades, it seems that Tehran has been built with a lack of continuity, not taking sufficient time to digest the sudden changes of urbanisation. Ali Madanipour (1998: 257) states that the city needs '*a degree of maturity, to counterbalance the constant flow of change*'.

However, Tehran's aesthetic, infrastructural, environmental, architectural, urban and spatial behavioural shortcomings are partly the reflection of its citizens' behavioural qualities; as well as lacking clear spatial policies. An ill sense of belonging, distinct spatial segregation between social classes with scarce opportunities for strong community involvement and the politicisation of many social activities means more struggles over the *right to* this *city*.

Experience has shown that Iranians have been socially and collectively proactive and helpful to one another in low times (e.g. The Iraq–Iran War in the 80s, the Bam Earthquake 2003 or the recent Ahar Earthquake in August 2010). There truly can be felt at many times a sense of altruism, humanism, compassion, diligence and national pride. These along with other qualities are ingredients for a potentially successful society; however for it to be just and for spatial behaviour as a portrayal of prevailing justice other supplementary elements are needed.

For urban citizens to feel at home, they need to acquire a sense of belonging to the city. The formation of community based organisations, (Non-profit organisations (NPOs), voluntary and charity groups) and their coalition could be a step forward. These could vary from statutory agencies with responsibility for community development or private sector organisations interested in community engagement, organisations run by and for communities of interest within the city of Tehran. While that being said, through regeneration and gentrification many local neighbourhoods such as Nazi Abad or Jannat Abaad have lost their once strong community identity over the past few years. By facilitating community participation, education and development, community based organisations could rejuvenate the true collective spirits of people in the capital and these neighbourhoods.

Further to NPOs, different unions and interdisciplinary involvements also nurture the relationship between individuals in society. Healthy microcosms of human activities in scattered public spaces in Tehran (As with the *Boustan parks*) along with appropriate pedestrianisation can support spatial interaction and participation.

As a variable dependent on time, the call for Tehran's spatial justice remains open-ended. In its very initial stages, the process for fulfilling spatial justice remains slow due to many premature decisions taken by officials who have been in and out of office, sometimes with good intentions and at other times self-centred actions. Though the prospect is promising a full time close observation through the eyes of many professionals is required for this ill child of the city to grow. Through this process, the city's multi-layered skin must be studied under the microscope by economists, urban planners and designers, architects, environmentalists, literary scholars, artists etc. However challenging it may be to abide by a clear plan, one lesson is clear: that when spatial justice is imprinted into the built environment, it is challenging to erase and 'if our urban world has been imagined and made, then it can be re-imagined and re-made' (Harvey 2003).

References

Harvey D (1978) The urban process under capitalism: a framework for analysis. Int J Urban Reg Res 2:101–131

Harvey D (1993) Social justice, postmodernism and the city. Int J Urban Reg Res 16:588–601

Harvey D (2003) Debates and developments. Right to the city. Int J Urban Reg Res 27(4):939–941
Lefebvre H, Kofman E (1996) Writings on cities. Wiley, Hoboken
Logan W (2000) Hanoi: biography of a city. University of Washington Press, Washington
Madanipour A (1998) Tehran: the making of a metropolis. Wiley, Chichester
Madanipour A (2006) Urban planning and development in Tehran. Cities 23(6):433–438
Mitchell D (2003) The right to the city: social justice and the fight for public space. The Guilford Press, New York
Soja EW (2010) Seeking spatial justice. University of Minnesota Press, Minneapolis
Soja EW, Kanai JM (2007) The urbanization of the world. In: Burdett R, Sudjic D (eds) The endless city. Phaidon, New York, pp 54–69
Tehran on the path of growth and justice Booklet, May 2012. The Municipality of Tehran Press, Tehran
The Javanmardaan Boustan booklet for the opening night, July 2012. The Municipality of Tehran Press, Tehran
Young IM (1990) Justice and the politics of difference. Princeton University Press, Princeton
Young I (1991) Justice and the politics of difference. Princeton University Press, Princeton

Chapter 8
Citizens as Drivers for Urban Change: Citizens' Exhibition as Participatory Tool. The Case of Hashtgerd New Town

Sabine Schröder and Jenny Schmithals

Abstract This chapter presents the participative-visual method 'citizens' exhibition' as a tool to foster public dialogue and discussion on energy and resource efficient behaviour and urban development in the New Town of Hashtgerd in Iran. The basic idea of a citizens' exhibition is to publicly present the attitudes, goals and motivations of different people in an exhibition in order to spur public discussion and promote understanding among citizens and different stakeholders concerning an issue. A citizens' exhibition is based on interviews with different actors. Pictures of the interviewees together with interview excerpts are arranged on posters, thus presenting a new and vivid view on the topic. The citizens' exhibition described in this chapter was developed in the framework of the German-Iranian research project 'Young Cities—developing energy-efficient urban fabric in the Tehran-Karaj region' dealing with the question of which role New Towns can play as a planning strategy of sustainable urban development. For a more attractive development of New Towns such as Hashtgerd the involvement of citizens in issues of urban development and planning processes is essential. In order to find out about the points of view of Hashtgerd New Town's residents on climate change, energy consumption patterns and obstacles as well as aspirations and needs for the development of Hashtgerd New Town, a survey was conducted among Hashtgerd citizens. From this survey a citizens' exhibition was developed displaying the most important and interesting views. The chapter describes the process of this citizens' exhibition as well as the interview results in detail.

Keywords Citizens' exhibition · Participation · Energy efficiency · Hashtgerd new town · Energy-efficient behaviour

S. Schröder (✉) · J. Schmithals
nexus Institute for Cooperation Management and Interdisciplinary Research, Berlin, Germany
e-mail: schroeder@nexusinstitut.de

J. Schmithals
e-mail: schmithals@nexusinstitut.de

F.F. Arefian and S.H.I. Moeini (eds.), *Urban Change in Iran*,
The Urban Book Series, DOI 10.1007/978-3-319-26115-7_8

Abbreviations

BMBF Bundesministerium für Bildung und Forschung (Federal Ministry of Education and Research)
BMU Bundesministerium für Umwelt (Federal Ministry for the Environment, Nature Conservation and Nuclear Safety)
IPCC International Panel on Climate Change
NT New Town
NTDC New Town Development Corporation

8.1 Introduction

Already in 2007 the reports of the International Panel on Climate Change (IPCC) had clarified that Climate Change is caused by anthropogenic actions and that counteractions are imperative if we want to limit the consequences of Climate Change already being experienced in different parts of the world to an extent that our society can cope with (cf. BMU, IPCC, BMBF 2007). At the same time, urbanisation rates are growing and approximately 80 % of worldwide greenhouse gas emissions are currently produced in cities. The building and housing as well as the transport sector are among the highest contributors in the emission of greenhouse gases. Therefore, the consequences of climate change can only be considerably reduced if we change the ways in which we build, move and live in cities significantly and sustainably (www.youngcities.org). This includes massive changes in individual energy consumption behaviour towards more sustainable energy consumption but also new concepts for energy-efficient urban planning and building which are accepted and valued by the people.

The research project 'Young Cities—Developing Urban Energy Efficiency in the Tehran-Karaj Region', which is funded under the programme 'Future Megacities' by the German Federal Ministry of Education and Research (BMBF), focuses on the energy efficient development of New Towns (NTs), with emphasis on the New Town Hashtgerd. The town is located in the growth corridor to the west of the expanding mega city of Tehran. Measured by its geographical size, it is the largest of the new urban settlements.

According to Walk and Schröder (2009) the discussion on the reasons and consequences of climate change is taking place at different levels today, but is only marginally reaching people. Attempts to change people's consumption behaviour in order to adapt to and mitigate the consequences of climate change are not reasonable without considering whether the climate change matters to people or not, which aspects of the topic are crucial to them and why, and which aspects are hindering or supporting the change of energy consumption behaviour in a given context. It is therefore necessary to analyse these questions to be able to adapt the design of the Young Cities project to the needs and lifestyles of the inhabitants of

Hashtgerd New Town and to develop urban concepts that support energy-efficient and ecologically sustainable lifestyles.

Therefore, as part of the Young Cities research project, an activating survey was carried out in order to investigate views and attitudes of the citizens of Hashtgerd New Town on climate change and energy efficient buildings, their energy consumption behaviour and whether or not they are prepared to save energy within the framework of the existing structures of the New Town of Hashtgerd, as the possibilities of saving energy or using it efficiently depend on the given urban structures, which can hinder or support sustainable lifestyles. A citizens' exhibition was developed from this survey in order to present the results of the study to a broader public and to start a thinking process and a dialogue on energy-efficient lifestyles in Hashtgerd that should put the voices of the inhabitants at the centre. It was to serve as a starting point for a further participation process, in which other participation mechanisms such as planning cells, future workshops or focus groups could bring about further solutions by citizens to improve living conditions in Hashtgerd New Town and thus trigger a change process.

8.2 The Participative Instrument: Citizens' Exhibition

Approaches based on participation, communication and education are essential to create and increase an awareness of the consequences of climate change and mitigation and adaptation strategies among affected stakeholders, and to persuade them to change their consumption behaviour. In order to promote an active engagement with the issue, it is particularly important that opportunities and spaces are created where existing views, attitudes and living circumstances can be publicly shown, reflected upon, and discussed.

The citizens' exhibition opens up these opportunities, as it presents attitudes, goals and points of views of stakeholders—such as residents of a neighbourhood, the municipal administration, private investors—in the form of excerpts from interviews together with photographs of people and their environment, in a public exhibition and thus linking participative elements with aesthetic components, giving a visual impression and showing experiences of various individuals, players and group representatives and their personal perspectives, attitudes and desires. However, the statements on the exhibition posters should not be mistaken as being representative research results. They are the statements of individuals that can be contradictory or provoking. The citizens' exhibition can thus spur a public dialogue or discussion, but also promote understanding different views concerning the selected issue. At the same time it can contribute to a stronger identification of participants with their group, the goals of their activities, and their neighbourhood, while also increasing the understanding of the viewpoints of other interest groups (Schophaus and Dienel 2002).

The concept of the citizens' exhibition was first developed as part of qualitative social research, dedicated to the analysis of the housing and living conditions in neighbourhoods and residential areas that are confronted with urban or social changes or are fighting social problems (cf. Legewie 2003). It was found that such a neighbourhood-based exhibition inspired some residents to actively be engaged in addressing the grievances raised there. Therefore, the method proved to have a good potential to activate people. From then on, this aspect was therefore emphasised in the method. Furthermore, the citizens' exhibition goes beyond researching a topic towards a more participative approach that seeks to spur a communication process based on the views of different actors. It was developed in the 1990s when participation practices were increasingly established in Germany (cf. Böhm et al. 2008). While other participative methods like the Citizens' Jury, Open Space Technology or Planning for Real developed in that time, are 'text-oriented and employ visual elements only in the form of cards that visualise textual results or in the form of technical models' (ibid.) and are mostly summarised in a written report, often without a greater public response, the citizens' exhibition as a participative method goes beyond this approach by using visual elements of exhibition to achieve a greater public response and spur a public dialogue (cf. ibid.). The major strength of the approach is the combination of visual and social research elements to bring through a certain topic with more emotional and aesthetic power than social research normally does, and its potential to motivate people to take action. The objectives of the citizens' exhibition are therefore threefold: (1) fostering participation by showing the views of different actors in a public exhibition, (2) motivating people to act, and (3) starting a communication process (Fig. 8.1).

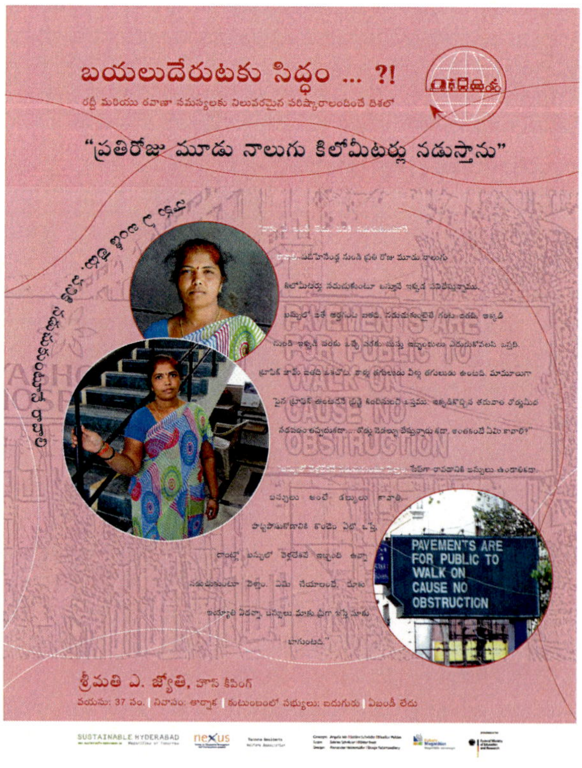

Fig. 8.1 Example of a citizens exhibition poster from the project 'sustainable Hyderabad' (Authors/nexus Institute. Reproduced by permission of nexus Institute)

The citizens' exhibition consists of several steps: conducting interviews with different stakeholders on a given topic, taking pictures of the interviewee and their environment, selecting the most important, short and characteristic statements of interviewees, designing exhibition posters with the excerpts based on interviews, taking photographs and selected personal information relating to the topic (e.g. age, profession, place of living) from the interviewee, and holding the exhibition.

The opening of the exhibition plays a key role for the instrument. All participating interest groups can be invited to come together, along with interested members of the public. For the opening of the exhibition the venue should be a local one. In addition to its public impact, another very important aspect of the exhibition opening is the opportunity it provides for initiating dialogue within and between the interest groups.

8.3 The Citizens' Exhibition 'Young Cities—Developing Urban Energy Efficiency: A View from Hashtgerd Citizens

8.3.1 Context and Goals

The Islamic Republic of Iran has experienced a massive population growth during the last decades. Its population has risen from 33.7 million in 1976 (Aghajanian 1995) to more than 78 million in 2012 (US Census Bureau). This development went alongside a rapid urbanisation process, following the global trend. Throughout this period the megacity of Tehran experienced drastic growth and grew in all directions along traffic arteries (cf. Madanipour 2005). As many other countries all over the world, Iran reacted with the strategy to build New Towns to relieve the rapidly growing agglomerations. Smaller cities are considered to be more manageable and offering a higher quality of life. The goal of the establishment of New Towns is to create living environments that are healthier than bursting megacities, while simultaneously bringing their growth under control (cf. ibid.). Although with the strategy of building New Towns an attempt was made to manage growth better and create better living conditions, this approach was not always evaluated positively. Negative consequences include the loss of green, natural areas, intensified social segregation and a slow development in many New Towns towards a functional, independent city. In many cases, the new cities lack necessary infrastructures and sufficient public transport, as well as jobs, for a considerable period of time. New Towns often attract a poor population. However, they can also attract middle-class citizens who can buy houses they would not be able to afford in larger cities (cf. Ghalehnoee and Diab 2005).

The planning of the New Town Hashtgerd was intended to relieve the fast growth of Tehran as well as the city of Karaj to its west. The planning for Hashtgerd New Town began in 1990. It was located north of the existing Old Hashtgerd,

30 km to the west of Karaj. Although the city was initially planned for 500,000 residents, it has today an estimated population of 20,000. Hashtgerd New Town is considered unattractive, due to a lack of shopping and cultural amenities and infrastructure. Most residents of Hashtgerd NT do not work within the city, but commute to Hashtgerd Old Town, Karaj or Tehran (Schröder et al. 2013).

The research project 'Young Cities—Developing Urban Energy Efficiency in the Tehran-Karaj Region', which is funded under the programme 'Future Megacities' by the German Federal Ministry of Education and Research, focuses on the energy efficient development of New Towns, with emphasis on the New Town Hashtgerd. The purpose of the project was to demonstrate and test energy-efficient building solutions. Among these is a 35 ha mixed-use residential area in the south of Hashtgerd New Town planned for approximately 8000 inhabitants. This settlement shall act as a showcase for energy-efficient and resource-saving urban planning and design that minimises technical complexities and costs and builds on the regional traditional knowledge of energy-efficient and resilient settlements. These also include four other pilot projects, all showcases for energy-efficient architecture, among them a five-storey apartment building completed in 2010, a Life Centre, which provides workshop space for vocational training in the building sector, an office building, and a variety of residential designs for the settlement (www.youngcities.org).

The citizens' exhibition described in this chapter was developed as part of the subproject 'Awareness Raising' within the Young Cities research project. The aims of the subproject were to find attitudes of the citizens of Hashtgerd on environmental issues and climate change as well as energy consumption patterns, and on that basis to raise awareness and sensitise people about sustainable energy consumption and energy-efficient buildings. Furthermore, insights would be obtained into the needs and suggestions of Hashtgerd residents regarding the urban infrastructure and the environment and how they hindered or fostered sustainable behaviour. This information would be incorporated into the work of the Young Cities project in order to integrate citizens' participation into the urban planning process of the project. In contrast to Germany, where there is a mandate in the urban planning process to inform citizens about plans and give them the opportunity to comment on them to be considered by the planners, there is no such obligation in Iran. Thus, it was the aim of this subproject to apply the German model and adapt it to the Iranian context and thus start a participative process to integrate the citizens' views into plans for the New Town of Hashtgerd.

In order to design attractive sustainable building solutions for people, it is essential to know, if and how sustainability matters to people and therefore, in what ways energy-efficient buildings can be a marketing asset and how it can improve current living conditions in Hashtgerd New Town. The survey was conducted as a basis for the citizens' exhibition whilst the citizens' exhibition itself gave a first insight into these questions.

The citizens' exhibition 'Young Cities—Developing Urban Energy Efficiency: A View from Hashtgerd Citizens' presented attitudes and knowledge of Hashtgerd residents about climate protection and energy-efficiency. The focus was on attitudes towards climate change, patterns of energy consumption and potential obstacles to energy-efficient behaviour within the urban or building environment, as well as examining people's contentment with the current infrastructure conditions in Hashtgerd New Town. The method of a citizens' exhibition was chosen, because it was recognised as a good opportunity to communicate the survey results to the Iranian partners and the inhabitants of Hashtgerd in a way which was convincing for the Iranian project partners. To begin, there was much scepticism about the necessity to take into account social and cultural aspects within building projects, and also about the chosen qualitative and participative approach. As an exhibition is a well-known instrument to present topics, the citizens' exhibition promised to be a good method as a starting point for a communication and a deeper participation process.

8.3.2 Conducting the Interviews

The survey that served as the basis for the citizens' exhibition was conducted in 2009 with 60 residents of Hashtgerd New Town by two Iranian interviewers. The interviews were officially registered and could, as a result, only be carried out in the presence of a government 'neighbourhood guardian' (moral guardian) (for this and the following description of the citizens' exhibition in Hashtgerd New Town, its results and adaptation, if not mentioned otherwise see also Schmithals and Poor-Rahim (2013), Schröder et al. (2013)).

The interviews were semi-structured with mostly open as well as closed questions and were led face-to face in public spaces or at the doorstep. Most interviewees answered willingly and were ready to talk about their living conditions even beyond the posed questions (especially when talking to women). The interviewees were to represent an average cross-section of the general population. To achieve this, a balanced proportion of gender and age groups was pursued. Of the 60 respondents, 29 were men and 31 women aged between 15 and 71 years. As the interviews were conducted during the day, it is likely that housewives, pensioners, the unemployed and students are slightly over-represented. The following personal data of the interviewees were recorded: age, education level, occupation, gender and place of residence within Hashtgerd New Town, Further socio-demographic information could not be requested, according to orders of the New Town Development Corporation (NTDC). According to the data collected, the interviewees represented a cross-section of the middle and lower middle class population living in Hashtgerd NT. Unfortunately, and in contrast to other examples of citizens' exhibitions, it was not allowed to record the interviews, nor to take photos of interviewees.

The presence of the guardians, who accompanied the interviewers, did not prove to be disturbing. The interviewees did not seem to be unsettled by their presence. They answered just as openly and extensively as did the interviewees in the few cases in which the guardian was not present.[1] For example, they critically commented on the structural condition of their houses and apartments.

8.3.3 Insights into Interview Results

The interviews were divided into several main parts: attitudes towards and knowledge about climate change and environmental issues and sources of information about these issues, energy consumption patterns regarding transport and on the household level, Hashtgerd's transport and social infrastructure as well as attitudes towards energy-efficient buildings. The answers to each topic showed different interesting results.

Regarding attitudes towards climate change, the survey showed that a majority of respondents feel threatened by the consequences of climate change and almost three fourth think that humans are responsible for climate change. By far the most important channel for information on issues of climate change and environment is the television.

One interviewee pointed out the familiar problem that the climate change is too abstract and not visible enough to the people to motivate individuals to commit themselves to reduce energy consumption, and also showed with his statement how the issue could also be connected to Muslim religion:

> I think climate change is somehow not sufficiently specific. Iranians do not step on a piece of bread lying on the ground, as they believe it is a gift of God. They treat things that are visible with caution. But climate change is not sufficiently visible and specific.
> (A 15 year female)[2]

Therefore, it is important to communicate to the public how climate change and its consequences affect their daily lives and what they can do to, for example, reduce energy consumption in their daily routines to make the issue more tangible and understandable.

Most respondents consider counteraction to climate change as being mainly, or at least partially, the responsibility of the state. Although many people stated that cooperation between citizens and the state is necessary, too: the state has the duty to create the necessary conditions for energy-efficient behaviour (e.g. inform the citizens), provide infrastructure, (e.g. adequate public transport infrastructure), and adopt and enforce binding laws. For example, several respondents demanded better information for the public about ways to counteract climate change, more cooperation with the people and integration into decisions in this regard. The need for binding legislation was emphasised by several respondents.

---

[1]In some cases, female interviewees invited the interviewers inside. As men cannot enter these places, these interviews were conducted without a neighborhood guardian.

[2]The citations mentioned in this article are part of the citizens' exhibition.

> In the industrialised countries the state was always the pioneer. You cannot protect the environment only on your own initiative. The state has to provide the framework for that. It will not help, for example, if every household separates its garbage, when it is not collected separately.
> (A 71 year male)

Interestingly, none of the respondents believed that nothing should be done about climate change, although this might have to do with socially desirable answering. All respondents thought that the government, citizens or both had the responsibility to take action against climate change.

The interviews showed that energy saving in the household was an issue, e.g. some make efforts not to heat all rooms during the winter, use energy-saving light bulbs or buy energy-efficient household appliances. However, the saving of energy is often hindered by the condition of the buildings. Therefore, energy saving measures often include measures to make up for the poor quality of the building. In the winter, the interviewees have to seal up their windows, the channels of the air cooling system and other leaking spots. Energy saving measures can thus be seen as taking necessary measures to achieve at least a certain degree of comfort in the apartment or house rather than being environmentally friendly behaviours.

The statements show that the efforts of the citizens of Hashtgerd New Town to save energy—be they for financial, comfort or environmental reasons—are impeded by poor construction conditions. Therefore, there is a big opportunity to save energy by avoiding holes and leakages and insulating the buildings and even more by building energy-efficient buildings. Most interviewees had heard about energy-efficient buildings, but most of them only associated double-glazed insulated windows and insulated exterior walls with the term energy-efficient building. Technology-oriented solutions for energy-efficient buildings were not mentioned and may not be well known. Also, there were differences of opinions on the feasibility of energy-efficient buildings in Hashtgerd. There were voices calling for better laws and better law enforcement regarding energy-efficient buildings and those who thought that energy-efficient buildings were too expensive—a thought that the Young Cities project is trying to prove wrong by using low-technology solutions.

Two-thirds of those interviewed lived in a household owning a car. However, nearly all expressed their wish for improved public transportation, such as Metro lines to Karaj and Tehran, reliable bus schedules, increased frequency of buses and more comfort in public transportation. Then, travel to these cities would be faster and more tolerable than it currently is and possibly more people would use public transportation which would result in less individual traffic and less energy consumption. In winter, especially, it is hardly tolerable to make these journeys, as buses are not heated:

> I would like to use buses if I wouldn't have to drag a blanket on the bus in winter.
> (A 26 year female)

Furthermore, Hashtgerd New Town, according to the respondents, lacks social, recreational and shopping facilities, forcing inhabitants to commute to

Old-Hashtgerd, or Karaj and Tehran which leads to longer travel distances and increases energy consumption. This shows that resource-saving behaviour is also hindered by the condition of transport and social and recreational infrastructures in Hashtgerd New Town.

The potential to conserve water was viewed more optimistically. For a large percentage of those interviewed, water conservation was an important issue, as water is a scarce resource in many parts of Iran and people try to save water in different ways. Seemingly, washing carpets and cars less often are seen as major measures for saving water:

> Certainly, we clean our carpets only once a year and wash our car only once a month – at home!
> (A 26 year female)

One interviewee doubted that saving water was even compatible with the Iranian lifestyle:

> I save everywhere, but saving water is not possible. I think that saving water would not fit well with the Iranian lifestyle!
> (A 34 year female)

The interviews with inhabitants of Hashtgerd also showed that sustainable or energy-saving behaviour is hindered by the condition of urban infrastructure, such as poor public transport, which leads to more use of private cars and social infrastructure. The existing building and infrastructure conditions not only make it difficult to save energy for the inhabitants of Hashtgerd, but also contribute in more complicated and uncomfortable living conditions. Energy-efficient urban structures, architecture and infrastructure could therefore not only reduce the consumption of energy but also massively improve the comfort of living for the future inhabitants of Hashtgerd or the 'Shahr-e-Javan Community' within Hashtgerd New Town.

The survey showed that energy consumption behaviour and the potential to reduce energy consumption is closely linked to urban planning and architecture. The citizens' exhibition therefore promised to be a good opportunity to bring this relation and needs for change regarding the living conditions in Hashtgerd New Town to the public through the voices of the inhabitants, in order to provoke a thinking and communication process on these topics and ideally lead to action on the side of planners, authorities or citizens.

8.3.4 Adaptation of the Method to the Iranian Context

The citizens' exhibition has so far mainly been applied in Germany, but there are also examples from other countries and different cultural backgrounds, such as the citizens' exhibition 'Ready to Move…?!' developed in the framework of an Indo-German research project in 2009, and the citizens' exhibition '¡Granada Limpia! Basura—Voces—Fotos', developed in the framework of German development cooperation in South America. However, the development and application of the method in Iran and in an Islamic context necessitated various adaptations for the method.

For cultural reasons, it was not possible to take pictures of the interviewees, so that no portraits of the interviewees, which are usually essential for the citizens' exhibition, could be taken. Also, it was not allowed to record the interviews, so that the interviewers had to closely take notes, nor was it allowed to take personal information from the interviewees, which are usually shown on the posters and also serve to invite the interviewees to the opening of the exhibition.

Therefore, the concisely noted quotations were sorted according to topics such as positions on climate change and energy, attitudes towards energy-efficient buildings, the potential to save water and the potential to save energy, together with the sex and age of the interviewees who were cited and presented on posters together with pictures relating to the topic of the posters, if possible from Hashtgerd New Town. Such images and individual quotations together with some personal information were combined on the posters. Furthermore, an introductory poster presented the objectives and goals of the Young Cities project and the subproject Awareness Raising as well as the background of the survey.

The citizens' exhibition was prepared bilingual presenting quotations in English and Persian languages in order to make them understandable in the Iranian context, but also for non-Persian speaking visitors. The following two pictures give an impression of the posters of the citizens' exhibition (Fig. 8.2).

8.3.5 The Opening of the Exhibition

The citizens' exhibition 'Young Cities—Developing Urban Energy Efficiency: A View from Hashtgerd Citizens' opened on the event of the inauguration of one of the pilot projects of the Young Cities project, 'The New Quality Building' in July 2010 in Hashtgerd New Town. The exhibition posters were presented in a prominent location in the entry area to the new building, which had to be passed on the way to the opening event. Unfortunately and unusually for citizens' exhibitions it was not possible to invite the interviewees to the exhibition opening, as it had not been allowed to take their personal data. Guests of the exhibition opening included, in addition to the German and Iranian project partners, other representatives of the Building and Housing Research Centre and the New Towns Development Corporation, top-ranking politicians as well as members of the Hashtgerd New Town municipal administration. Representatives of print media and television were also present. Due to a last-minute cancellation of the Construction Minister the inauguration event was postponed by half an hour, which caused many of those present to use this time to view the exhibition and discuss it (Fig. 8.3).

Fig. 8.2 *Right* example poster from the Young Cities citizens' exhibition. *Left* example poster from the Young Cities citizens' exhibition (Authors/nexus Institute. Reproduced by permission of nexus Institute)

Fig. 8.3 Picture of the opening of the citizens' exhibition (Courtesy of Leslie Quitzow 2009)

8.4 Conclusion

Generally speaking, a public exhibition is not expected to have a direct effect on the activation of stakeholders, although it is possible in exceptional cases. Activation is a process and can best be realised by offering many of these and similar activating events or instruments. However, the citizens' exhibition offers the opportunity to visualise opinions, attitudes and knowledge of the involved actors to the public and thus start a process of dialogue and opinion formation. The local opening can be considered a success. The Iranian project partners received it positively, despite the fact that the partners initially did not consider interviews to be crucial to the project and that the interviews could only be carried out under strict constraints.

The citizens' exhibition 'Young Cities—Developing Urban Energy Efficiency: A View from Hashtgerd Citizens' provided a forum to visualise the views of the inhabitants of Hashtgerd New Town to the public. It also showed that sustainable consumption behaviour is massively impeded by poor construction conditions, public transport and social infrastructure of Hashtgerd New Town. The existing building and infrastructure condition not only make it difficult to save energy for the inhabitants of Hashtgerd, but also contribute to more complicated and uncomfortable living conditions. Energy-efficient urban structures, architecture and infrastructure could therefore not only reduce the consumption of energy but also massively improve the comfort of living for the future inhabitants of Hashtgerd or the 'Shahre Javan Community' within Hashtgerd New Town. Therefore, the citizens' exhibition calls for changes of urban planning and architecture towards an urban planning that supports sustainable lifestyles.

The presentation of the citizens' exhibition was intended to pave the way for further participative methods. Different participative measures to follow up the process had been introduced to the Iranian partners for implementation. Due to much scepticism or low interest in participative measures on the side of Iranian partners, smaller participative approaches, namely focus group discussion (small groups that lead a structured and moderated discussion focused on a specific topic) were suggested, but finally failed to be implemented due to the impossibility of obtaining an official permission. However, this shows that the citizens' exhibition is a very well suited participative and communicative method which can even be implemented in difficult situations or when participation is normally seen with scepticism. There are fewer obstacles than with other participatory approaches, as interviews and exhibitions are widely spread and accepted concepts. Also, the method could be adapted to the Iranian cultural context through some minor conceptual changes to the method and can thus be recommended for implementation in other cultural contexts, e.g. in Muslim nations. This citizens' exhibition shall therefore be further spread in Hashtgerd New Town and through the internet to enhance its activating impact and its potential to support change in urban planning and living conditions for more sustainable lifestyles.

References

Aghajanian A (1995) A new direction in population policy and family planning in the Islamic Republic of Iran. United Nations Asia-Pacific Popul J 10(1):3–20. Available from http://www.un.org/Depts/escap/pop/journal/v10n1a1.htm. Accessed 15 Aug 2012

BMU (Federal Ministry for the Environment, Nature Conservation and Nuclear Safety), IPCC (German Coordination Office for the Intergovernmental Panel on Climate Change) BMBF (Federal Ministry of Education and Research) (2007): 4. Sachstandsbericht (AR4) des IPCC (2007) über Klimaänderungen. Teil III—Verminderung des Klimawandels. http://www.bmu.de/files/pdfs/allgemein/application/pdf/ipcc_teil3_kurzfassung.pdf. Accessed 15 Aug 2012

Böhm B, Legewie H, Dienel H-L (2008) The citizens' exhibition: a combination of socio-scientific, participative and artistic elements. FQS (Online) 9(2), Art. 33. Available from http://www.qualitative-research.net/index.php/fqs/article/view/380

Ghalehnoee M, Diab Y (2005) Application of an indicator system of sustainability in new towns; case study: Iranian new towns. In: International conference on New Towns. Full articles. Tehran, pp 291–304

Legewie H (2003) Feldforschung in historischen Stadtquartieren: Von Berlin nach Florenz und zurück. Ein Erfahrungsbericht. In: Legewie H (ed) Erzählungen und Bilder der Stadt: Lebensqualität und Tourismus in historischen Vierteln von Florenz und Berlin. Technische Universität Berlin, Berlin: University Library

Madanipour A (2005) Why do we build new towns? Experiences of Britain and Iran. In: International conference on New Towns. Full articles. Tehran, pp 69–82

Schmithals J, Poor-Rahim N (2013) Die Bürgerausstellung in einem islamischen Kontext: Young Cities developing urban energy efficiency. Tehran Karaj. In: Keppler D, Böhm B, Dienel H-L (eds) Die Bürgerausstellung—Die Perspektive von Bürgern und Bürgerinnen als Gegenstand qualitativer Sozialforschung und praktischer Beteiligung. oekom, Munich

Schophaus M, Dienel H-L (2002) Bürgerausstellung—ein neues Beteiligungsverfahren für die Stadtplanung. Forschungsjournal Neue Soziale Bewegungen 15(2):90–99

Schröder et al (2013) Energy consumption patterns and attitudes towards climate change in Hashtgerd New Town. Young Cities Research Briefs 01. University Press of the Technische Universität Berlin, Berlin

US Census Bureau. Demographic overview—custom region—Iran (Online). Available from http://www.census.gov/population/international/data/idb/region.php. Accessed 19 Aug 2012

Walk H, Schröder S (2009) Low emission lifestyles in megacities. Communication and participation strategies in Hyderabad (Online). Working paper July 2009. Institute of European Studies, UC Berkeley. Available from http://www.escholarship.org/uc/item/4vw3775g. Accessed Aug 2012

www.youngcities.org Introduction to Young Cities research (Internet), 19 Apr 2011. Available from www.youngcities.org/intro.html. Accessed 15 Aug 2012

Part III
Urban Continuity or Discontinuity

Chapter 9
Tehran, the Scene of Modernity in the Pahlavi Dynasty: Modernisation and Urbanisation Processes 1925–1979

Azadeh Mashayekhi

Abstract In 1925 the Pahlavi monarchy (1925–1979) embarked on a project to create a new modern capital city while modernising Iranian society and creating a modern nation-state. This chapter presents a framework to analyse the transformation of the changing urban form of Tehran during this time. Two elements of this framework stand out; First, the concept of modernity, which emphasises the condition of continuous change in the city. The second element is surveying political geography of Tehran during this period and traces its link to social and urban transformation of the city. Hence this chapter uses several events and particular national development plans of Pahlavi dynasty since the end of the First World War till Islamic Revolution in Iran, in order to understand the impact of different kinds of policies and interactions with different kinds of place on shaping the form and life of Tehran. Ultimately this chapter illustrates the ways in which national modernisation projects of the Pahlavi dynasty produce a particular kind of city while at the same time produce a particular form of urban society.

Keywords Urban form · Modernisation process · Urban transformation · Political subject

Abbreviation

TCP Tehran comprehensive plan

A. Mashayekhi (✉)
Faculty of Architecture, Spatial Planning and Strategies, Delft University of Technology, Delft, The Netherland
e-mail: A.mashayekhi@tudelft.nl

F.F. Arefian and S.H.I. Moeini (eds.), *Urban Change in Iran*,
The Urban Book Series, DOI 10.1007/978-3-319-26115-7_9

9.1 Introduction

Throughout the twentieth century Iranian cities have undergone processes of modernisation in successive political regimes that have left their traces in the physical and social form of the city. This chapter attempts to investigate underlying dynamics of modernisation processes and their effect on the urban form of Tehran, focusing on the post-WWI period until the Islamic Revolution (1925–1979). This period coincides with the constitution of the royal state of Iran founded by Reza Shah Pahlavi in 1925 and ruled by him and his son Mohammad Reza Shah till 1979. This period is a critical time in Iranian modern history. It overlaps with struggles for building a modern nation, industrialisation, and nationalisation of the oil industry that led to a CIA engineered coup d'état in 1953, and finally the 1979 Islamic Revolution and the fall of Mohammad Reza Shah Pahlavi. Tehran throughout the rule of the Pahlavi dynasty transformed from a walled city with 200,000 inhabitants to a metropolis of five million, with half a million dwellers living around the city in new suburbs (Atlas of Tehran Metropolis 2005). Throughout this period, successive plans and urban interventions have brought about massive changes in the urban form of Tehran. These plans and interventions were key contributors to the modernisation process of the capital. Hence this chapter seeks to explore the complex relationship between planning interventions, urban form and social organisation in Tehran. The aim is to trace a contiguous process of building a modern city and society, by revealing the serial character of these processes and the way modernity, coming from the outside, was manifested in different forms through plans and visions of the city and its culture.

This period (1925–1979) contains two critical phases of modern planning history for the entire Middle East (Habibi 1997; Madanipour 1998). The first phase begins with the end of the First World War and ends with the beginning of the geopolitical tensions between the USSR and the United States (the Cold War). During this period, colonial rules were established in many countries in the region and modernisation and secularisation began to expand into Muslim territories. The second phase (the Cold War period) is a time when the US became progressively more committed to the stability and security of the Middle East and to the suppression of Soviet Communist influence in the region (Boyer 2010). Each of these phases had brought particular political intentions for planning intervention, although throughout both periods, one main strategy was consistent for Iran and many other countries in the region; the making of a modern national space and models for a new citizen (Cinar 2007; Schayegh 2012). In Iran, a series of national development plans and modernisation projects were prepared to achieve the strategy of modern space and society building, which led to a considerable transformation of urban centres and urban society (Madanipour 2010).

The objective here is to understand two major phases of urban modernity and the ways in which a number of particular national development plans and projects in these two periods produced particular forms of the city and urban society (Habibi 2009). In this sense, Tehran's urban form transformation was directly influenced by these national modernisation projects, such as building a centralised state, Westernisation, land reform, abolishing traditional feudalism, and industrialisation. All these development projects were highly influenced by a European, and later American, model of modernisation. This chapter traces the link between the implementation of these modernity projects and the remaking of the capital city and its citizens. Moreover. It seeks to demonstrate that Tehran's urban transformation process is not only an evolutionary transition from old to new, but it is a juxtaposition of historical layers of modernity and how one modernity project builds on previous projects to redefine the space of the city. Ultimately, it seeks to contribute to planning history and the history of urban transformation in Tehran.

The next section seeks to explain the application of the term 'modernity' as a framework, to study the spatial effects of modernisation, followed by describing how the Pahlavi kings (father and son) set the agenda for the political consolidation of the modern state and the means to achieve this end. The next section will analyse the two conditions under which European and American visions of the modern city and state took hold and where elements of these visions came from. Finally, the outcome of these visions and plans—new infrastructures, new public buildings, new industries, and new housing areas—will be discussed. The chapter concludes with an interpretation and assessment of the socio-spatial changes that characterise each of these phases of 'modernity' in Tehran.

9.2 Modernity as a Framework

> All people, things, institutions and environments that are innovative and avant-garde at one historical moment will become backward and obsolescent in the next. Even in the most highly developed parts of the world, all individuals, groups and communities are under constant relentless pressure to reconstruct themselves; if they stop to rest, to be what they are, they will be swept away. (Berman 1982: 78)

The term modernity is used here as a framework for the analysis of transformation of urban form in Tehran. Here we use the term modernity to emphasise the condition of continuous change and innovation in cities (Berman 1982). Since the early nineteenth century in Iran, different concepts of modernity and what constitutes being modern have been at the centre of social and political discourse. As the Iranian case illustrates, regardless of how it is defined, the idea of modernity has an immense transformative power in the ongoing formation of a social–political and spatial order of the country, and the shaping and transformation of cities and urban life. In such a case, 'it is much more meaningful to study the specific meanings modernity takes locally and how such meanings take form in different modernisation projects' (Cinar 2007: 152). The significance of the historical processes of the creation of this local modernity is in fact particularly rich in the case of a country like Iran, whose founding ideology in the sense of nation building and industrialisation is so engaged with the ideal of Western modernism.

This study adopts a pluralised concept of modernity (Said 1978; Berman 1982; Rabinow 1989; Mitchell 1988; Eisenstadt 2000; Taylor 1999; Robinson 2006; Cinar 2007) as an analytical method to trace different modernisation[1] projects within this historical process. The pluralised concept of modernity as Paul Rabinow argues, looks at multiple ideas of modernity that have been produced and reproduced at particular historical times and places (Rabinow 1989). Each modernity consists of explicit projects or plans for the transformation (modernisation) of society. These plans use the city as a medium or instrument to shape the modernity project. Consequently urban plans mainly become a tool and are associated with different social modernisation projects (Mitchell 1988; Rabinow 1989; Cinar 2007). This chapter will adhere to the pluralist approach to study Tehran's modernisation processes—as it opens up the possibility to discuss modernity in non-western contexts—asserting that modernity is a project with distinct phases, of which two periods are discussed here: 'Modernity.1', the First Pahlavi (1925–1945) and 'Modernity.2', the Second Pahlavi (1945–1979).

Each of these two 'modernity' periods shows specific historic visions and plans to control change, that involved new functions and institutions associated with urban space as well as new ways of life. In other words, these two phases of modernity show the ways in which modernisation projects are related to the transformation of the city and urban society. However, it is also important to make clear how the implementation of modernisation plans and projects are constrained by conditions produced by previous plans during previous periods (Boyer 1983; Harvey 2006; Lefebvre 2003). Therefore, the answer to a number of questions is sought here: first, how the juxtaposition of historical layers of modernity projects through time determines the city we encounter today. Second, how each modernity project has been constrained in relation to past projects.

9.3 The Pahlavi Period (1925–1979): Strategies, Plans and Projects

This section describes very briefly the political geography of Tehran during the Pahlavi dynasty (1925–1979), and a number of key modernisation projects and national development plans that had direct influence on socio-spatial transformation in Tehran in this particular period.

In 1925, Reza Shah, with the help of a British-backed coup, and the support of the modern middle class and secular nationalist elite,[2] established the Pahlavi

[1]By modernisation we mean a new way of organising life and society; transformations of materialist civilisation- improvements in sewage system, pluming, transportation, and material efficiencies.

[2]Secular nationalist elite mainly composed of military leaders succeeded in securing Iran's interior from any remaining domestic and foreign threats, prominent bureaucrats, influential writers and journalists mostly educated in Europe.

dynasty. As Reza Shah came to power the country was thinly populated—its 1.6 million square kilometer of territory had only 12 million inhabitants, with a large nomadic population, and only 21 % of the populace living in urban centres (Abrahamian 2008). In fact Reza Shah took over a country that was lacking central bureaucracy, and had a weak economic system dominated by agriculture (Cronin 2003). Therefore, the Pahlavi dynasty aimed to construct a national public sphere where Iran's national identity could be performed and displayed for a global audience. Tehran became the showcase for Pahlavi's modern ideas and projects. Two main national plans and projects proposed by Reza Shah resulted in projects for reorganising and modernising Tehran. These two plans are: administrative reform (modernising the structure of the state and the economy), and building a new network of infrastructure throughout the whole country (industrialisation), which transformed Iran into an urbanised country. These two national projects were carried out very soon after the Shah consolidated his power and none of them had primacy over the other. The Shah proposed seven ministries located in Tehran (commerce; post and telegraph; endowments; education; road; industry; and agriculture), and modelled on 19th century Europe which all grew to become a substantial bureaucracy. The Trans-Iranian railway project was proposed as part of the national plans for modern infrastructure to connect Tehran with different parts of the country. The modern railway network aimed to link Tehran with Torkaman Port on the Caspian Sea in the north and the Shahpur Port on the Persian Gulf in the south which were both industrial centres based on textile and oil. Ultimately, these plans reinforced the unity of the territory and entered Iran into an industrialisation process.

The modernity projects of the First Pahlavi were halted by the advent of WWII and the abdication of Reza Shah. During his rule, developments not only accelerated with great speed, but also became part of a much more grandiose project. The establishment of a modern state and central rule, brought law and order, discipline, central authority and modern amenities—universities, schools, transport, communication, and entertainment. However, WWII and a series of political events of the decade after the war influenced these development processes considerably. In the first decade after WWII, Iran was characterised by national and regional tensions as a result of the nationalisation of the oil industry by Prime Minister Mohammed Mosaddeq in 1951 that prompted his removal in a CIA-instigated coup two years later (Katouzian 2009). After the fall of Mosaddeq's in 1953, Reza Shah's son, Mohammed Reza, consolidated his autocratic rule—and followed the legacy of his father; modernisation, industrialisation, and developing the country's infrastructural facilities (Banani 1961; Madanipour 1998). By the early 1950s his son was keen to accelerate the pre-war policies of his father (Reza Shah) with a new system led through planning. His approach was different from his father's; he had strong tendencies towards Americans since they had helped him to regain his power and he was in favour of 'building big' as part of the national modernisation policy. Two major catalysts promoted and controlled modernisation projects in this period: 'Truman's Point Four Programme' and the 'Plan Organisation' in Iran (Karimi 2009).

The Point Four Programme was a technical assistance programme for 'developing countries' announced by the United States in 1949. Following WWII, the United States found itself in a Cold War struggle against the USSR. President Truman's administration came up with the idea for a technical assistance programme as a means to win the 'hearts and minds' of the developing world. By sharing US know-how in various fields, especially agriculture, industry, and health, officials could help 'third world' nations on the development path to raise the standard of living, and show that democracy and capitalism could provide the welfare of the individual (Schayegh 2008; Karimi 2009). Americans justified their interventions in Iran and other developing countries in the region as humanitarian help to modernise these countries (Karimi 2009; Provoost 2006). Nevertheless the policy was determined to furnish Iran with Western values and ideas, technologies, and commodities in an effort to integrate Iran into the world capitalist economy (Karimi 2009).

Only less than a decade after the war, the idea of reconstructing and developing Iranian cities through government initiatives gained fresh ground in the country (Daftary 1971). A new government agency was shaped to supervise Iran's planning development activities called the 'Plan Organisation' (*Sāzmān-e Barnāma*) (Iranica 2014). The chief function of the Plan Organisation was to design development plans and to coordinate and supervise their execution, which was the responsibility of the respective ministries. The Plan Organisation was financed by a share of oil revenues, and borrowed extensively from the World Bank and the 'Point Four Programme' (Karimi 2009; Madanipour 2010). Over nearly 30 years, the Plan Organisation provided five series of seven-year national development plans, each of which containing projects such as building dams and roads, and improving the public health system and rural life. These extensive projects of industrialisation, urban development, and national health and education were largely financed by oil income. The technical bureau of the organisation employed a great number of Western advisors, to help with the ambitious development projects of the time. Among them at the level of urban development projects were Louis Kahn, Constantinos Doxiadis, Victor Gruen, and Kenzo Tange (Madanipour 2010; Emami 2011).

Between 1956 and 1966 the Plan Organisation's industrialisation strategy had huge impact on urban centres especially in Tehran. Among them was the 'land reform' policy that had considerable impact on the urbanisation process of the whole country. This government initiated property redistribution aimed at abolishing the traditional feudalism and enhancing the industrialisation process (Bayat 2010; Vahdat Zad 2013). Turning feudal landowners into sharecropping farmers and the peasantry into smallholders released roughly three million landless peasants from the countryside (Bayat 2010; Vahdat Zad 2013). This led to the massive rural-urban migration, freeing up a surplus labour force to be absorbed in the industrial sector, Tehran's population grew from half a million in 1939 to 1.5 million in 1956, and 2.7 million in 1966 (Madanipour 2010). This massive urban growth continued during 1960s and 1970s, accompanied by remarkable economic growth as a result of skyrocketing oil revenues.

What follows is an examination of how national strategies, development plans, administrative reform, national infrastructure, the emergence of the 'Plan Organisation', and land reforms led to radical urban change and the emergence of new patterns, institutions, and functions in Tehran in this period, among other things demonstrating how the concrete socio-spatial effects of these plans shaped Tehran's urban character and the spatial distribution of its population.

9.4 Modernity Projects for a New Capital: Tehran in Transition

In the 1930s, a number of projects were initiated for Tehran as a result of the trans-national railway project and administration reform plan. One of the most important ones was a new urban plan for Tehran called the 'Street-Widening Act of 1933'. The plan aimed for a new urban order in the capital and to improve the quality of the built environment. It called for a French-style modern street plan that imposed a degree of geometric regularity, with monuments as focal points of the street system (Fig. 9.1) (Cronin 2003). This plan proposed new urban forms and functions alongside newly created avenues, previously unknown in the Iranian tradition. In 1937, the 'Street-Widening Act of 1933' was officially inaugurated. The plan not only proposed new boulevards but also the widening of old streets. Soon, new public buildings, and educational and commercial activities spread out along newly built avenues mainly towards the west and north. North-south streets and the new east-west Shah Reza Avenue in the north enjoyed the highest standards of infrastructure and services (Fig. 9.1). Almost all new urban administrative activities were located in the north and many new industrial buildings in the south. Therefore, due to the growing bureaucracy and new industries, two different types of urban activities began to form in different parts of the city.

The new administrative district was designed and executed on two sites– the former walled Qajar Royal District and the former military parade ground– were converted into the centre of government with new ministries and public buildings, such as the Ministry of Justice, the Ministry of Foreign Affairs, the Ministry of War, the National Museum, the first national Iranian bank, the national radio station and an office for post and telegraph. These processes were partially governed by the fact that the north was becoming the area where wealth and power concentrated, in contrast to the south, which lacked such a concentration. This dichotomy became even more obvious when the southern parts of the city turned into an industrial hub with a modern railway station, military airport, tobacco factory, cement factory, textile factory, the Tehran Silo (cement factory), and power plant. All these new industries were located in the south, southeast and southwest of the city mainly alongside roads that connect Tehran to other cities and provinces (Fig. 9.2).

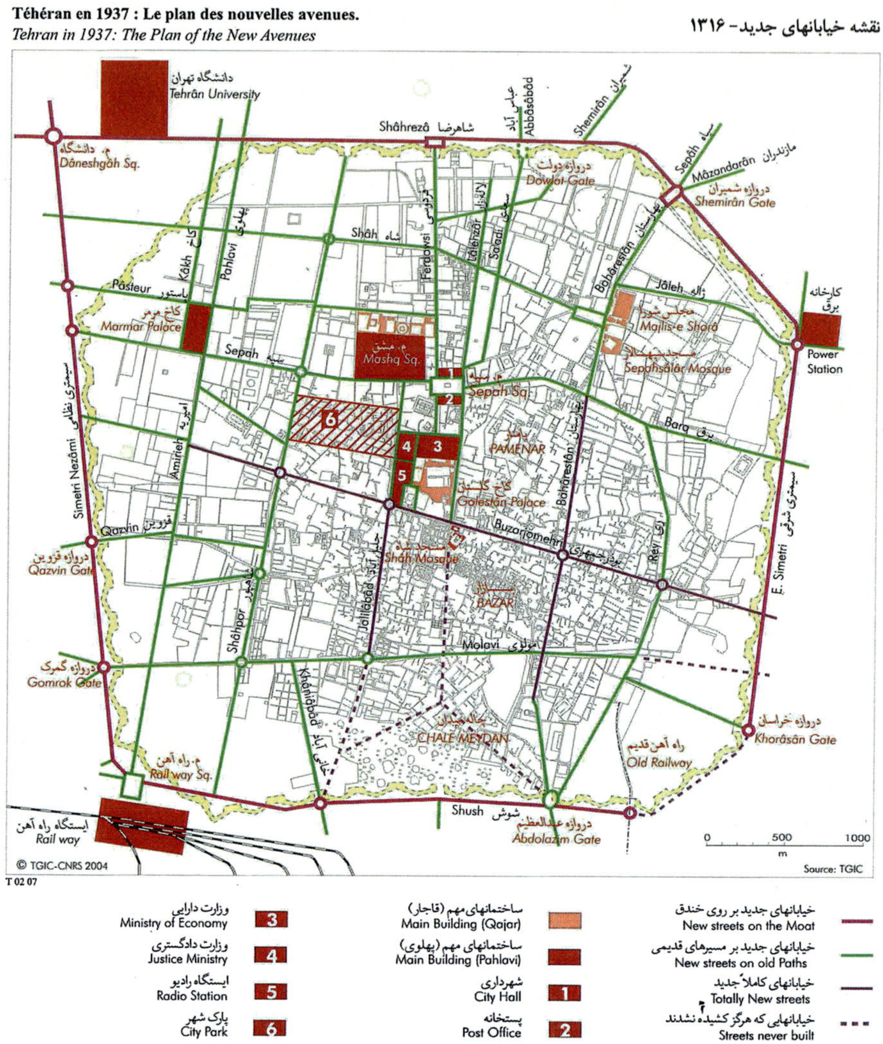

Fig. 9.1 Tehran map in 1937, presents the plan of the new grid of streets and emergence of new institutions (Atlas of Tehran Metropolis 2005). Reproduced with permission of Tehran Municipality ICT Organisation

By the end of the first Pahlavi regime in 1941, Tehran covered an area of around 46 km^2, which was 2.5 times bigger than during the Qajar period. The population grew from 200,000 to nearly one million in only sixteen years (Atlas of Tehran Metropolis 2005). Streets and avenues covered 1.8 km^2 or 9 % of the whole area of the city. For Tehran, the two decades that followed the end of the first Pahlavi were

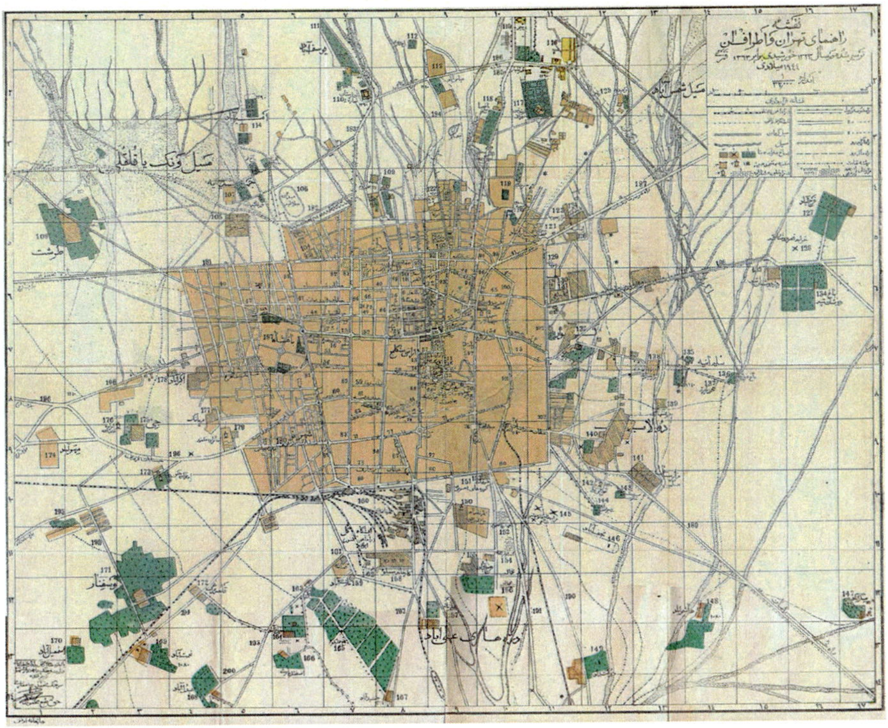

Fig. 9.2 Tehran in mid-1940s, the map shows the new railway station located in the south and development of new industries around the station and urban growth towards north and west parts (Archive of Sahab, published online: shahrsazi.tehran.ir)

characterised by rapid growth of the city mainly towards the north and west, the expansion of commercial activities along the main avenues, high density in the city centre and widespread unemployment in poor areas, along with continuous migration of working classes (Madanipour 1998). By the late 1950s and early 1960s, housing costs and availability were major problems for the city, leading to a proliferation of new neighbourhoods outside the city, mainly in the north and north-west, where professional salaried groups settled.

Between 1950 and 1960, the population of Tehran doubled to two million. The middle class population growth was by far the quickest among other social groups. The main reasons for this development were a rise in state employment, due mainly to the state's growing revenues after Iran's nationalisation of oil in 1953, and a laissez-faire economic policy that made loans more readily available. The Plan Organisation as part of the second (1956–1961) and third national development plans (1963–67) prepared two major development projects for the growing population of Tehran that led to a new form of the city and a new modern urban lifestyle (Encyclopaedia Iranica 2011). These plans are the Karaj Dam project (1958–1961)

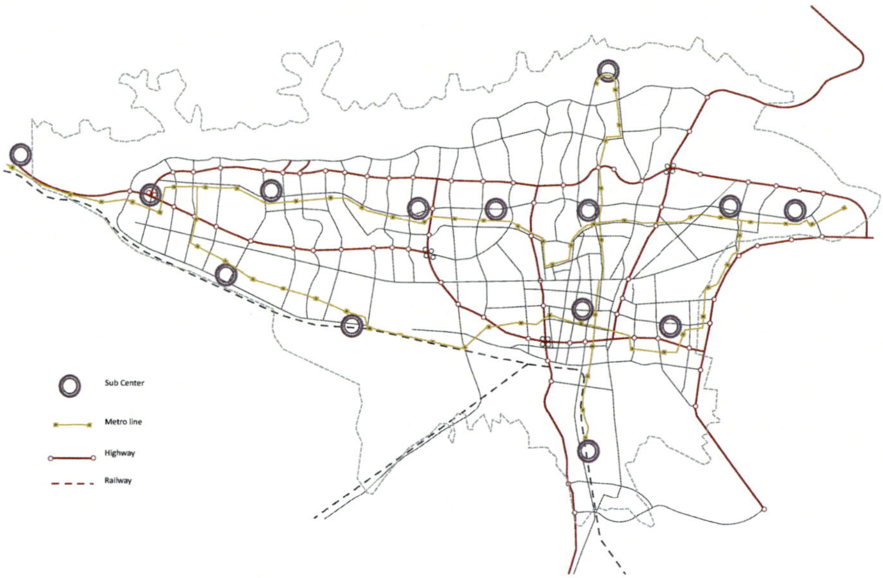

Fig. 9.3 The overall scheme of the first Tehran master plan 1964 (Author)

and the Tehran Comprehensive Plan (TCP) (1964–1968). Both projects represented a vision to create an economically prosperous city and society with high living standards. The Karaj Dam project responded to a demand for electricity and a surge in popular expectations for mass consumer goods and higher living standards that began after WWII (ibid.). While the TCP was designed to shape Tehran's modernisation, accommodate the growing population of Tehran's middle class, and provide facilities for a 'modern-lifestyle'.

The Karaj Dam project was part of the Shah's bias towards development and 'building big'. He got the idea for dam projects after he visited dams in Japan and India (Schayegh 2012). 'In his eyes, dams expedited development, were a powerful symbol of his commitment to modernise the country.' (ibid.: 637). Thus in the 1950s, the detailed plan for construction of the Karaj Dam[3] was prepared as one of the grand modernity projects of the Shah to shape the modern life in Tehran. The US policy of 'war on communism' and developing Western mass consumerism

[3]Karaj Dam is located 60 km northwest of Tehran and 23 km north of Karaj. It was constructed on the Karaj River. The initial studies for Karaj Dam took 22 years completing in 1956 when formal proceedings began and Morrison-Knudsen Co. constructed the dam in the period from 1958 to 1961. The Karaj Dam was built as a multi-purpose dam to provide tap water for Tehran alongside agricultural water for Karaj. It supplies the irrigation demand of over 50,000 ha (120,000 acres) of farmlands near Karaj. The power plant has been connected to the national electricity grid for over 46 years and has a capacity of 90 MW. The lake behind the dam is a tourist attraction while being a natural habitat for rainbow trout. With its sailing and water-skiing facilities, the dam is a popular weekend summer resort.

in the third world countries contributed considerably to the realisation of this project. The Plan Organisation began the construction of the massive hydro-electrical dam on the Karaj River 60 km north of Tehran in 1958 with the help of American financing and technical experts (ibid.). Though the demand for clean water dated back to the late nineteenth century in Iran, the electricity during the post-WWII years was not simply a necessity but a sign of modernity, essential in industrial production and a prerequisite not only for lighting but also for consuming a new range of modern products such as refrigerators, televisions, and washing machines (ibid.). In 1961, Karaj Dam's electricity flooded Tehran, and became a source of power and water for the city.

By the early 1960s, more than 30 % of Iran's urban population were housed in Tehran (Amirahmadi 1986). In 1964 for the task of designing a 25-year comprehensive growth plan for Tehran, the architectural firm Abdulaziz Farmanfarmaian[4] was selected by the 'Planning Organisation' to collaborate with the American partner Victor Gruen,[5] who was known as the 'Mall Maker' and the designer who gave architectural shape to American consumerism (Hardwick 2004). The Tehran Comprehensive Plan (TCP) that was approved in 1968 became the most important post-war planning document for the city (Madanipour 2010). The plan proposed decentralisation, relocation and displacement of the poor and had the vision that Tehran as a metropolis in 1991 would consist of ten urban centres, each with half a million inhabitants, separated by large green areas and linked with a network of highways (Fig. 9.3). A total length of 150 km of highways was planned for the city, and a transportation network design in the east-west direction to support the linear growth of the city towards the west (Emami 2011). This plan intended to accommodate commercial markets and facilitate the movement of automobiles. Tehran was to become 'not a diagram of power, but a machine for profit' (Karimi 2009).

The TCP represented a vision evidently striving to provide the facilities of a 'modern life-style'. The TCP design for residential prototypes for different income groups reveals the vision of producing this new lifestyle in the city. The typical high-income and low-income communities were designed for Tehran, one with large plots arranged around open spaces and accessed by highways and the other composed of an orthogonal grid with small plots (Fig. 9.4). Single families were living in luxurious apartment buildings with swimming pools and tennis courts enhancing the image of high-income neighbourhoods. For all these families, automobiles, televisions, refrigerators, and gardens would be essential elements of their lifestyle. However, the obvious contrast between the design of low-income and high-income neighbourhoods indicates that the TCP designed the entire city for an income group whose life-style would appeal to other income groups. The

[4]Farmanfarmaian was from an aristocratic family and a graduate of the Ecole de Beaux Arts in Paris. He had established the first 'consultant architects firm' in the country after his return from Paris in 1950.

[5]Victor Gruen was the Austrian-American architect and planner based in Los Angeles. By the early 1960s, He had shifted from designing malls to urban planning, and proposed an ideal diagram for a metropolitan region, resembling Ebenezer Howard's Garden City.

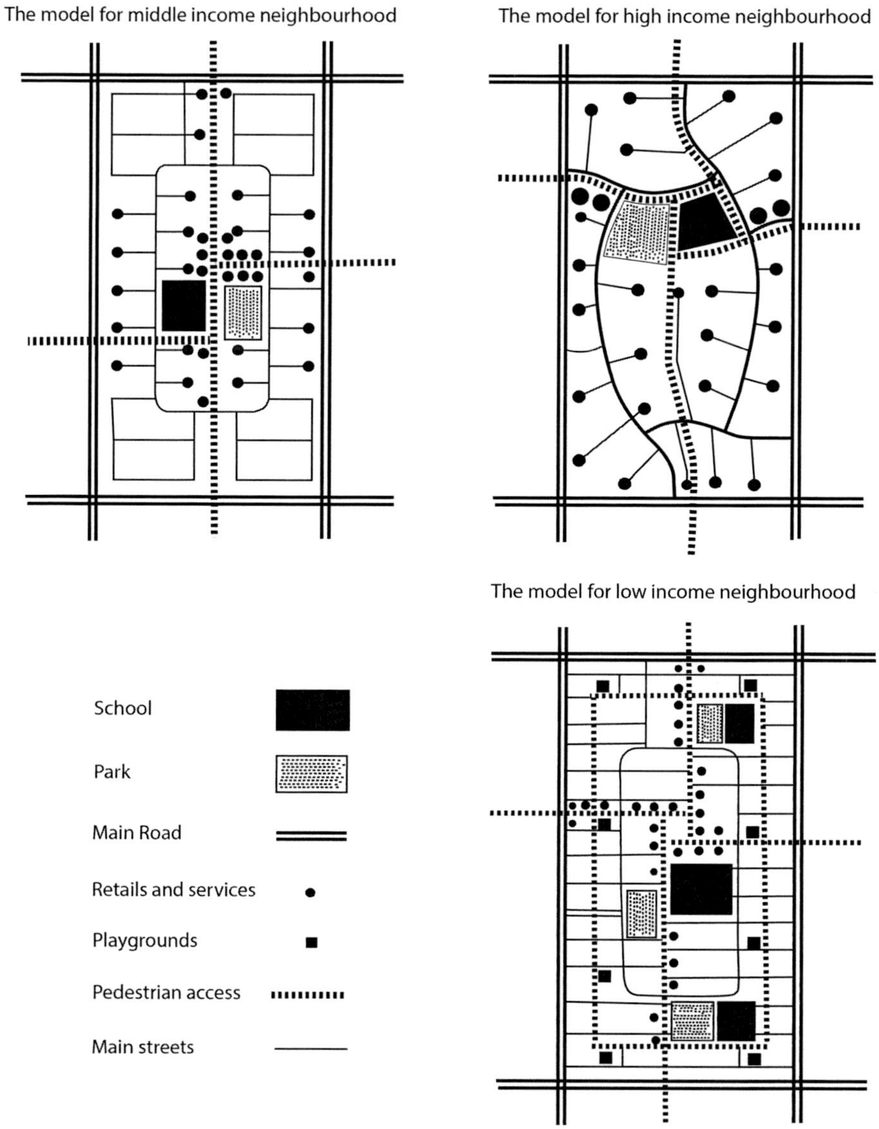

Fig. 9.4 The schematic plans for residential neighbourhoods for high income and middle (*top*) and low income (*bottom*) communities (Author)

assumption of TCP was that Tehran will be more socially mobile, and that a high quality urban lifestyle will help to raise the aspirations of people from poor backgrounds to get top jobs and earn high incomes. Ultimately, Tehran would be the utopia of lower income groups regardless of the deficiencies they face in the city (Emami 2011).

By the end of the Shah's reign in 1979, Gruen's plan for Tehran was only partially realised (Karimi 2009). Part of the planned transportation network and a number of large-scale housing projects were realised or under construction by the time of the 1979 Islamic revolution (Karimi 2009; Madanipour 1998). Throughout the 1960s and 1970s the government and the private sector started building housing for different income groups following the planned segregation of income groups. There are major examples of these projects, including Kuy-e-Kan and Lavizan, both developed by the Ministry of Housing and Development in north-west and north-cast Tehran. Kuy-e-Kan has some 1000 three-bedroom flats, built with subsidies for 475,000 government employees with medium incomes. Lavizan is for 45,000 low-to-medium income government employees. Shahrak-e-Gharb was developed in a north-west suburb of Tehran by the Pahlavi Foundation, largely financed by the private sector for middle-to-upper income groups. Abbasabad was largely developed by the private sector as a high-income residential-business district north of the city centre (Amirahmadi and Kiafar 1987).

A major consequence of these development projects and policies was widening inequality gaps and increasing differences among Tehran's residential districts (Amirahmadi 1986, Hourcade 1974). By the end of Pahlavi's dynasty the population of Tehran was 4.5 million and between 42 and 50 % of the population had housing problems (Amirahmadi and Kiafar 1987). By the mid-1970s 82.7 % of all national companies were registered in the capital, and the city had 52.9 % of all banking units in the country. Meanwhile 40 % of total movie tickets were sold in the cinemas of Tehran. Additionally, one in every ten Tehranis had a car, compared to one in every ninety in other cities. The heavy concentration of wealth, industry, and services in Tehran was the key to rapid growth and uneven distribution of population at the national level.

9.5 The Social Conditions of the New Modernity

The modernisation and Westernisation before the Pahlavi dynasty was confined exclusively to the upper class and royal family circles in Tehran. Under the Pahlavis modernisation became official state policy and a way of life for different social classes. The administrative reform of the first Pahlavi and massive industrialisation during the second Pahlavi led to the emergence of various social groups of middle class and industrial working class in the city. By the end of the 1970s large nomadic populations became urban dwellers, and 47 % of Iran's population in 1976 lived in urban centres. The continuous urban transformation during the Pahlavi rule (1925–1979) changed the social organisation of Tehran's public spaces and the spatial distribution of population. The implementation of modernisation plans led to the construction of spaces of new bourgeoisie along boulevards and highways, which transformed the old traditional districts of former tribal chiefs and aristocrats in the centre of the city to working class districts and shaped affluent suburbs in the northern parts. The construction of dual spaces of modernity (urban middle class versus industrial working class) led to a complex layering of engineering urban spaces and a way of life within Tehran. The increasing number of new government offices, industrial factories, and commercial centres in Tehran reinforced this duality (Amirahmadi and Kiafar 1987).

The population of urban elite and urban middle classes in particular, played an important role in Tehran, performing and displaying Iran's modern national identity by living a life that they saw as modern and western. In other words, what was imagined as the modern European was in fact the very local bourgeois culture of Paris. Observing the social-spatial condition of Tehran during first Pahlavi's modernity period, Annie Boyce sees the city not only as a place where 'shops predominated in the avenues (*Khiyabans*) … [and] the street front[s] of many buildings were decorated by protruding neoclassical columns.', but also as the 'Mecca of the dandies' (Boynce 1923; Karimi 2009: 104). She notes that 'Laleh Za*r* Avenue was a promenade for the perfumed youth of the city who were now dressed according to the 'national style'[6] (ibid.). In the writings of local authors, the dandies of Tehran in this period are referred to as *fokoli* [the man with the suit and bow tie], occasionally seen as comparable with the 19th century dandies of Paris or *flâneur* as Walter Benjamin called them (Milani 2004). Pamela Karimi emphasises the desire of the Tehrani middle-classes during the second Pahlavi period to imitate foreign modes of living and compared this group with the *petite bourgeoisie* in the late 19th century France (Karimi 2009). She emphasises mass consumerism culture and the desire for western commodities among Tehran's citizens and notes that 'if we look at the messages of Iranian advertising in the mid-20th century, Iranians were meant to follow the path of the west and like the European petit bourgeois, were encouraged to "live beyond their means".'

9.6 Conclusion

The fall of Mohammad Reza Shah as a result of the Islamic Revolution in 1979 is the proof of the failure of many of his government policies, and can be an example of what Berman calls 'the tragedy of development' (Emami 2011). The poor and middle classes poured into the streets of Tehran demanding democracy and social justice, among other things. Tehran became a historic scene of the revolution with massive demonstrations largely by marginalised Tehranis and squatters from the southern districts (Ehsani 1999). Both development plans of the father and son did not liberate the economic forces nor did they increase social mobility. However these two periods of modernity in many ways were different and specific to their time. Unlike Reza Shah's 1937 superimposition of a street network on the

[6]In 1929 Iranian men and women were banned from wearing their ethnic and traditional costumes and they had to follow the 'national style'. Based on this law men were required to wear Western-style suit with a new hat, designed based on the French military kepi, called Pahlavi hat. Women were banned to wear hijab and the police were arresting women who wore the veil.

traditional and organic neighbourhoods of the capital, Gruen's plan for Tehran is neither geometric nor saturated with conception of order and discipline. The street plan of Reza Shah was responding to nationalism and the power of modern state, with city squares where he built statues of himself and past national figures and ancient kings. The Gruen's plan and other projects like Karaj Dam from the Plan Organisation during the Second Pahlavi were not projects of power but machines for profit. The TCP's programme was to accommodate commercial markets and facilitate movement of automobiles to diminish distances in the city. Ultimately, we can simply trace the link between national modernisation projects and urban plans and their socio-spatial outcomes in these two periods.

This chapter has made efforts to grasp and define how particular modernity projects build on previous projects to produce new spaces and society during Pahlavi dynasty period. Regardless of how successful the Pahlavi modernisation project of Tehran was, I insist on the construction of two particular forms of modernity that are distinct and different from other modernity phases. In this chapter I have shown the ways the city acquired a new internal structure and, on the regional level, a form which is open to the countryside. Socially, we see drastic changes: the process of widening contradictions, and an emerging north-south division which did not exist before, and the arrival of two new urban social classes of considerable size: an urban middle class and an urban poor industrial working class. This trend intensified in the next period of modernity 1979 until today, but under different modernisation projects and dynamics, such as suburbanisation, and motorcar and mass consumption culture.

References

Abrahamian E (2008) A history of modern iran. Cambridge University Press, New York

Adele C, Hourcade B (eds) (1992) Teheran Capitale Bicentenaire. Institut Francais de Recherche en Iran, Paris and Tehran

Amirahmadi H, Kiafar A (1987) Tehran growth and contradiction. J Plan Educ Res 6(3):167–177

Banani, A (1961). The Modernisation of Iran. Stanford University Press, Stanford

Bahrambeygui H (1977) Tehran: an urban analysis. Shahab Books Institute, Tehran

Bayat A (2010) Tehran: the paradox city. New Left Rev 66:99–122

Berman M (1982) All that is solid melts into air: the experience of modernity. Simon & Schuster, New York

Bonnett A (2005) Occidentalism and plural modernities: or how Fukuzawa and Tagore invented the West. Environ Plan D Soc Space 23(4):505–525

Boyce A (1923) Chapters from the life of an American Woman in the Shah's capital. Presbyterian Hist Arch, Philadelphia

Boyer C (1983) Dreaming the rational city: the myth of American city planning 1890–1945. The MIT Press, Cambridge

Boyer C (2010) Adventures on Baghdad: constantinos A. Doxiadis, the Science of Ekistics and Cold War Politic [online]. Available from http://vimeo.com/17431716. Accessed May 2012

Çınar A (2007) The imagined community as Urban reality, The making of Ankara. In: Çınar A, Bender T (eds) Urban imaginaries, locating the modern city. University of Minnesota, Minneapolis

Cronin S (2003) Modernity, change and dictatorship in Iran: the new order and its opponents, 1927–29. Middle Eastern Stud 39(2):1–36
Daftary F (1971). Economic development and planning in Iran (1955–1967), Ph.D. thesis, University of California, Berkeley, pp. 329–507
Ehsani K (1999). Municipal matters: The urbanization of consciousness and political change in Tehran'. Middle East Rep (Merip) (212): 22–7
Eisenstadt SN (2000) Multiple modernities. Daedalus 129(1):1–29
Emami F (2011) Civic Visions, National Politics, and international designs: three proposals for a new urban center in Tehran (1966–76), master thesis, Massachusetts Institute of Technology
Galantay EY (1987) Islamic identity and the metropolis: Continuity and conflict. In: Saqqaf AY (ed) The middle east city: Ancient traditions confront a modern world. Paragon House, New York, pp 6–24
Habibi SM (1996) Tahavvol va Gostareshe Tehran dar Zamane Reza Shah, 1304–1320 (Transformation and development of Tehran during the Reza Shah's period, 1925–1941) (in Persian)
Habibi SM (1997) De la cite a la Ville. University of Tehran Press, Tehran
Habibi SM, Ahari Z, Emami F (2009). From demolishing fortifications to thoughts of highways: history of urban design in Tehran from 1930 till 1966. Soffeh Journal. 50(20):51–61
Habibi SM, Hourcade B (2005) Atlas of Tehran metropolis. Pardazesh va Barnamerizi-e Shahri Publication, Municipality of Tehran
Hardwick MJ (2004) Mall maker: Victor Gruen, Architect of an American Dream. University of Pennsylvania Press, Philadelphia
Harvey D (2006) The political economy of public space. In: Low S, Smith N (eds) The politics of public space. Routledge, New York, pp 17–34
Hourcade B (1974) Téhéran: évolution récente d'une métropole . In: Méditerranée, Deuxième série, Tome 16, pp 25–41
Hourcade B, Adle S (eds) (1996) Tehran, Payetakhte devist saleh (Tehran, the 200 years old capital), Tehran Engineering and Technical Consulting Organisation, Tehran (in Persian)
http://shahrsazi.tehran.ir/LinkClick.aspx?fileticket=zTGa4Kfiw5U%3d&tabid=456&portalid=0&mid=1362 [online]. Accessed May 2012
Iranica, 2014 http://www.iranicaonline.org/articles/barnama-rizi-planning [online]. Accessed May 2012
http://www.iranicaonline.org/articles/cities-iv [online]. Accessed May 2012
http://www.iranicaonline.org/articles/ebtehaj-abolhassan [online]. Accessed May 2012
Karimi P (2009) Transitions in domestic architecture and home culture in twentieth-century Iran, Thesis (Ph.D.), Massachusetts Institute of Technology
Katouzian S (1996) Tehran, capital city: 1786-1997, the re-invention of a metropolis. In: Petruccioli A, Ludovico M, Alemi M (eds), Environmental design: Iranian architecture in search for a new identity. IEDRC, Como, Italy, pp 34–45
Katouzian H (2009) The Persians: ancient. Yale University, Mediaeval and Modern Iran
Lefebvre H (2003) The urban revolution. University of Minnesota Press, Minneapolis
Madanipour A (1998) Tehran: the making of a metropolis. Wiley, New York
Madanipour A (2006) Urban planning and development in Tehran. Cities 23(6):433–438
Mandinapour A (2010) The limits of scientific planning: Doxiadis and the Tehran action plan. Plan Perspect 25(4):485–504
Milani A (2004) Lost wisdom: rethinking modernity in Iran. Mage Publishers, Washington, DC
Mitchell T (1988) Colonising Egypt. University of California Press, Berkley
Provoost M (2006) New towns on the Cold War frontier: how modern urban planning was exported as an instrument in the battle for the developing world. http://www.eurozine.com
Rabinow P (1989) French modern: norms and forms of the social environment. MIT Press, New York
Robinson J (2006) Cities between modernity and development. S Afr Geogr J 86(1):17–22
Said EW (1978) Orientalism. Penguin Group, London

Schayegh C (2012) Iran's Karaj Dam affair: emerging mass consumerism, the politics of promise, and the cold war in the Third World. Compar Stud Soc His 54(3):612–643

Seger M (1975) Strukturelemente der Stadt Teheran und das Modell der modernen orientalischen Stadt. Erdkunde 29:21–38

Taylor PJ (1999) Modernities: a geohistorical perspective. Polity, Cambridge

Vahdat Zad V (2013) Spatial discrimination in Tehran's modern urban planning 1906–1979. J Plan His 12(1):49–62 (originally published online 21 September 2012)

Chapter 10
From Utopia to Dystopia: Shushtar-e-No, Endeavour Towards Paradigmatic Shift

M. Reza Shirazi

Abstract In the 1970s, and to meet the needs of the growing urban population, the Iranian government was able to plan for large-scale investments in the housing sector, thanks to suddenly increasing oil revenues. For a while, the western approach to housing and dwelling, crystallised in the proliferating and internationally appreciated style of modern architecture, dominated Iran's new architecture and urban planning. However, a number of architects and urban planners resisted this dominance and tried to create a paradigmatic shift in the approach to housing and dwelling, focusing on the place-specific aspects of the context. The Shushtar-e-No project was an endeavour of this kind. This chapter addresses this paradigmatic shift, focusing on the case of Shushtar-e-No, a satellite city located 2 km from the old city of Shushtar and designed by Kamran Diba. After a short introduction to the political, social, and architectural context of the scheme, the chapter highlights how the architect's unique approach to the built environment promised a paradigmatic shift in the question of housing and dwelling, the aim of which was to 'synthesise' the two modes of tradition and modernity in quest of a 'local style', and to promote a 'social agenda'. Next, an investigation of the current environmental, social, and physical situation of the community will show its degeneration from the initial utopian image into a state of dystopia, which can be linked with both the initial architectural pre-suppositions and with later unexpected political incidents. Ultimately, using Foucauldian terminology, it will be concluded that Shushtar-e-No has transformed to a 'crisis community', a 'forgotten land', which represents a heterotopia *par excellence*.

Keywords Utopia · Dystopia · Paradigmatic shift · Shushtar-e-No · Community design

---

M.R. Shirazi (✉)
School of the Built Environment, Oxford Brookes University, Oxford, UK
e-mail: shirazi@brookes.ac.uk

F.F. Arefian and S.H.I. Moeini (eds.), *Urban Change in Iran*,
The Urban Book Series, DOI 10.1007/978-3-319-26115-7_10

10.1 Introduction

The plans and programmes for modernisation of Iran were established during the reign of Reza Shah (1921–1941), and followed by the young Mohammad Reza Shah (1941–1978), who fulfilled his father's dream of creating a massive state structure. The state bureaucracy underwent drastic expansion; a series of seven-year development plans was launched to achieve industrial prosperity. The results included a maximised growth rate and a substantial increase in national revenues. Social programmes enhanced the standard of social life in the fields of education, health, women's participation, the family and so on. All these efforts aimed to bring about a 'Great Civilisation', which aspired to be 'the fifth most industrial state in the world'. As Ervand Abrahamian spells it out, the claim of the time was that 'Iran was at the gates of the Great Civilisation; its future would be more glorious than its past…; its standard of living would soon surpass that of Europe; it would produce a way of life superior to both capitalism and communism…' (Abrahamian 2008: 131).

This was the context in which large-scale investment was committed to housing by the government, thanks to the sudden increase in oil revenues in the 1970s, necessitating short term construction activities to meet the needs of the growing urban population (Habibi 2008; Madanipour 2006). This goal could be achieved by 'prefabricated or industrial housing, which also sounded suitably in tune with an industrially-aspiring nation' (Diba 1980a: 38). This construction method, using imported materials and assembled on site, neglected local materials and labour, imposed a different life style, and ignored the inhabitants' cultural and social particularities. Although the trend was welcomed by the Western-oriented professionals of the time, some architects who were sufficiently aware of the socio-cultural problems originating from these kinds of housing and construction systems, resisted their logic and dominance, and tried to open up new place-specific perspectives to this issue. To understand these attempts at resistance, it will be helpful to put them into context through a brief retrospective overview.

The first key steps towards modernist architecture were taken under Reza Shah who tried to bring Iran into the modern world and stimulate modernisation (Abrahamian 1982). Additionally, there was growing attention to the pre-Islamic legitimation of the Kingdom in order to 'Persianise' Iranian culture along pre-Islamic lines (Kamali 1998). In this context, a number of Iranians trained in European universities such as Vartan Avanessian (1896–82), Paul Abkar (1908–70), Gabriel Guevrekian (1900–70), and Keyghobad Zafar (born 1910), 'introduced into Iran a new language that broke all links and continuity with the past' (Micara 1996: 54). On the other hand, European architects or archaeologists such as Maxim Siroux, André Godard and Arthur Pope tried to concretise the nationalist wishes of the government with regard to architectural buildings through 'a superficial repetition of architectural forms, elements and motifs, particularly those of the Achaemenid and Sassanid periods' (Mirmiran 2004: 39) in a nineteenth century Neoclassical European manner based on the 'application of symmetry, hierarchy and geometric forms' (Diba and Beheshti 2004: 32). The first line of thought was later intensified under Mohammad Reza Shah in both architectural and urban planning projects. For example, Aziz Farmanfarmaian provided the first Master Plan of Tehran in 1968, together with Victor Gruen Associates, in which he followed a Los Angelesian pattern (Habibi and Hourcade 2005) with an extensive focus on physical aspects and the unrelated transfer of ideas from the West 'without pursuing their social objectives' (Madanipour 1998: 208). In his architectural works as well, he was influenced by the universal principles of the International Style, as exemplified in projects such as the Azadi Sports Complex, Ministry of Agriculture, Saman Towers, and Mehr Abad Airport.

Although this paradigm dominated the professional sphere in the 1960s and 1970s, there was a growing tendency towards integrating tradition and modernism in a constructive way, to create an intermediate condition where: 'The primary values of architecture rather than the forms are brought into evidence, defining a more intellectual and abstract idea of tradition' (Micara 1996: 62), thus advocating a paradigmatic shift. This approach, which is observable in certain works of Houshang Seyhoun, was followed more systematically by leading figures such as Kamran Diba, Nader Ardalan, and Hossein Amanat, who suggested a mode of architecture which is not based on the direct employment of architectural forms and decorations, but on new interpretations of traditional primary values, a kind of 'modern regionally-inspired architecture' (Khan 2000: xxxi). This resisted the governing trend of vulgar modernism, and opened a new perspective in which tradition and modernity can combine to generate a new approach. As an attempt to generate a paradigmatic shift, to instil an overwhelming new paradigm, this could never achieve success. However, this was not due to any lack of theoretical grounding or capacity in its prominent figures, nor due to the reluctance of the architectural profession, but as the result of the radical political changes that took place in the wake of the Islamic Revolution of 1979. This led the burgeoning bottom-up paradigmatic shift to degrade into an ideological top-down call for an Islamic architecture and urbanism, an ambitious aspiration which has never been realised on the ground, nor even taken seriously by professionals.

As a prominent figure, Kamran Diba was deeply aware of the necessity of highlighting place-specific particularities and linking architecture to its socio-cultural context, and as a result designed some architectural and urban works which deserve more attention and contemplation. The Shushtar-e-No project is a work of this kind, in which the two factors of 'housing' and 'regionally-inspired architecture' come together in a creative and innovative way. This article intends to clarify how the Shushtar-e-No complex represented a call for a paradigmatic shift in the question of community and collective dwelling, by means of architectural and urban initiatives and innovations proposed by its architect. This utopian image, however, degraded into a disastrous condition after the project's half-realisation, to the extent that it is now more dystopian, portraying a 'crisis community' far removed from its original 'image'. To figure out the reasons behind this degradation, the role of architectural miscalculations and the subsequent unsystematic developments in housing policy will be discussed.

Prior to focusing on the architectural and urban characteristics of the work, it will be helpful to highlight some critical aspects of the architect's philosophy, which will help us to better understand his unique approach to designing this complex.

10.2 Synthesis of Tradition and Modernity Through the Architecture of Resistance

Diba's architecture is formed by a continuous endeavour to combine tradition and modernity, to create, as he explains, 'a dialectic and synthesis' between these two modes (Diba 1998: 87). He regards the indigenous architecture and culture of Iran as precious sources, but, at the same time, that an absolute emphasis on tradition may lead to a reactionary position and halt creativity. Therefore, a safer approach is to synthesise the two modes of tradition and modernity in a creative way. It is precisely this synthetic approach which makes Diba's contribution to the above-mentioned paradigmatic shift a very significant one.

For Diba, modernism is not merely a physical agenda addressing the appearance of the built environment, but rather a 'social programme' which focuses on human values and ideas (Javaherian and Diba 2005). However, he criticises the Western imposition of values as well as scientific, technological, and production methods in less-developed countries and argues that the 'Loss of independence and perhaps a perceived humiliating subjugation to Western ideas and values has brought about an identity crisis in Islamic societies' (Diba 2002: 119). The irony is, he asserts, that while sometimes aware of the problems of this phenomenon, all Islamic societies with different ideologies are building a 'western-style metropolitan life' and re-produce the same high-rise buildings, slums, and mega-transportation systems at the expense of social coherence and the environment. He states that this endeavour is more superficial than constructive, because 'while Islamic societies are arguing over the design of the buttons, the suit is being produced according to the Occidental model' (Diba 2002: 120). In this sense, Diba highlights the importance of the container, not the ornaments; he instigates a quest to provide urban enclaves which address the very cultural and social identity of society. In fact, the 'search for a local style' has always been Diba's desire and inclination (Diba 2010: 125), a goal which is to be achieved through simultaneous attention to the universal and the particular.

This belief in synthesising modes of tradition and modernity makes Diba sensitive to the overwhelming tendencies of the time, which have overemphasised the modern side. As Diba states, Shushtar-e-No project was a 'reaction' against prevailing trends in contemporary Iranian architecture, mainly influenced by modernism and the International Style, with their extensive inattention to the everyday world of the people and their traditional lifestyles based around the essential role of neighbourhoods (Diba 1998). This reaction could be understood as a 'peripheral movement' taking place in an era when progressive architecture and urban planning were essentially in thrall to the International Style and modernism. Thus, in this context of extreme modernisation and universalisation, Diba advocates resistance against the dominant architectural paradigm, to awaken policy makers and developers to its emerging dangers. This sense of resistance, for example, can be seen in Diba's opposition to the development plan for the Imam Reza Shrine in Mashhad, where the aim was to destroy the old historical context around the shrine to provide space for new developments and urban facilities. His argument was that urban life is organic in form, so cuts to the existing urban texture will simultaneously demolish urban life. His efforts, though unsuccessful, convey his resistance to vulgar modernist approaches to urban development (Diba 2010: 145–146).

In the case of Shushtar-e-No, this resistance played a vital role, with much effort put into convincing developers to go beyond the dominant presuppositions concerning mass housing production and encouraging them to refer to more indigenous concepts of dwelling and housing. Contrary to the earlier Imam Reza Shrine, he was this time successful in convincing the client and was able to incorporate his ideas in the body of the project. This enabled him to design a unique residential complex through a mixture of innovative envisioning and programmes.

10.3 Shushtar-e-No, A General View

In 1974, the agro-industrial complex Kesht-o Sanat-e Karoun decided to plan a residential and urban complex for its employees. In 1975 D.A.Z. Architects, headed by Kamran Diba, were commissioned to prepare plans. Shushtar New Town (Shushtar-e-No) was a new residential community, a satellite city containing houses, schools, a commercial centre, and a mosque, to house around 25–30,000 white- and blue-collar workers from the aforementioned company. It is adjacent to the city of Shushtar in Khuzestan, southern Iran, located to the north of the city, and only 2 km away from the city centre, divided from it by the Shatit River but connected via the Shadravan Bridge. The city of Shushtar is an ancient city which follows the traditional urban pattern typical of Iranian cities. The original urban fabric still preserved and observable from the time of its design, is compact and interwoven, with mud-brick as the predominant construction material (Diba 1986).

In Shushtar-e-No the spinal cord of a main axis extending from end to end of the complex serves as a connector through which all the major components, districts, and activities are linked together. On the eastern side, an open space provides a location where local farmers can sell their vegetables and other foodstuffs, and thus connects the hinterland activities to the urban fabric; a progressive idea which may hinder the extreme segregation of the city from its surrounding agricultural zones. A covered street bazaar culminates in a green public space at the heart of the central residential quarter. A commercial centre is located in the middle with spaces arranged around a courtyard. This neighbourhood has its own local bathhouse as well as a local mosque. The main axis is continued with the community and cultural centre. The Friday Mosque is located at a high point at the mid-point of the long

Fig. 10.1 Site plan of Shuhstar-e-No. Reproduced with permission of Kamran Diba

pedestrian boulevard within a man-made forest. The main axis continues with a pedestrian square and then a park, and finally culminates in the town square (Maidan-e Shahr), with dimensions of 100 × 100 m, which is connected to the old city via a pedestrian bridge and works as the principle urban space unifying the old and new towns (Javaherian and Diba 2005; Diba 1981) (Fig. 10.1).

Different types of residential units have been arranged around this spinal cord. Other urban facilities such as different types of schools are distributed in the area within an accessible distance. A main sports facility is located in the south-western area, a bus station in the south. Two bridges connect the complex to the old city. The construction of the project began in 1976, and all five phases were to be completed by 1985. However, due to the ensuing political transformation, the first phase was only partially built, representing some 650 homes, with the mosque, library, and some public spaces remaining uncompleted (Diba 1985).

A range of architectural innovations come together to make this project unique, addressing different aspects from environmental adaptability to social agenda, from the micro scale of residential units up to the macro scale of public spaces. One of the project's main distinctive features derives from the 'social agenda' Diba has envisaged for this complex. Diba's attention to public space and social interaction can be traced back to his interest in society and social sciences, which had concerned him to the extent that he studied sociology for one year at a post-graduate level (Diba 2010). He writes: 'I became fascinated by the laws that govern social systems. Understanding social situations not only introduced a new challenge and a new dimension to my work, it led to an anti-architectural attitude' (Diba 1981: 8). He had been intrigued by the social expression seen in the street life of his country, where groups idle on street corners, parade themselves in the course of shopping expeditions and interact. On this basis, he understands a built structure as a 'social event', 'an environment which multiplies and enhances the quality of interaction' (ibid.). A main task for an architect, he believes, as with a theatre director, is organising and setting the stage for certain human situations and providing users with the possibility of interaction. This tendency towards public architecture is so high that Diba has mainly designed public projects, and has always declined to take on commissions for villas and private houses, except in the case of one designed for a sculptor friend, Parviz Tanavoli (Diba 2010).

Diba's interest in public spaces is well illustrated in Shafagh Park (1966–69), his first architectural work in Iran. The central concept was a programme for 'a community centre' consisting of two small libraries for adults and children, a workshop for children, a community hall and offices dedicated to different welfare organisations, as a place for socialising and social participation (Diba 1981). Encouraging social interaction is also a key point in Shushtar-e-No, so that Diba talks about a 'social agenda' (Javaherian and Diba 2005) for the town. A main pedestrian and social 'spine' consisting of a series of paved squares, lush gardens, covered and shaded resting places, fountains and running water, which are lined with schools and bazaars, has been designed to stimulate socialisation. All the streets and passages culminate in this spine, encouraging people to socialise (Diba 1981). This scheme, similar to the pedestrian pathway of Jondi-Shapour, now Shahid Chamran University, stems from the idea of a 'human interaction-intensification program', that is, 'enhancing the quality and quantity of human interaction by means of physical, spatial organisation' (Diba, 3: 54).

A more private sense of public space has been provided in front of some houses, where the house meets the street, as 'extensions of houses where children play and parents chat' (Diba 1981: 186). This realm was to be a semi-private, semi-public one at the service of the inhabitants and neighbours. As Diba puts it, 'The basic attempt was to create a socio-physical entity conducive to collective interaction, togetherness and strong community ties' (ibid., 186).

The presence of public space is amplified by the neighbourhood plaza, where a shaded area is provided with shops and tea houses, and by the neighbourhood mosque in the middle of the residential quarter, which follows the traditional pattern in terms of its integration with the surroundings (Fig. 10.2). At the city scale, the main plaza (Maidan-e Shahr), a $100 \times 100\ \mathrm{m}^2$, was to be situated on the river bank, and connected to the old city through a pedestrian bridge. This unrealised Maidan, as the connector of the old and new city, was to house government offices, a hotel, apartments, arcade shops, and cultural activities such as cinemas.

Due to Shushtar's geographic location, the climate is tough and harsh, and the precipitation rate very low. To mitigate these severe conditions, the planning concept of the city is dense in pattern with deep and narrow brick-paved streets, so that the buildings provide shading for living and traffic spaces. This is significant when the existing hot climate necessitates natural cooling. Such cooling occurs when the heat rising from streets is cooled by breezes and also through surface water drainage. In the central shopping centre, deep arcades provide shaded

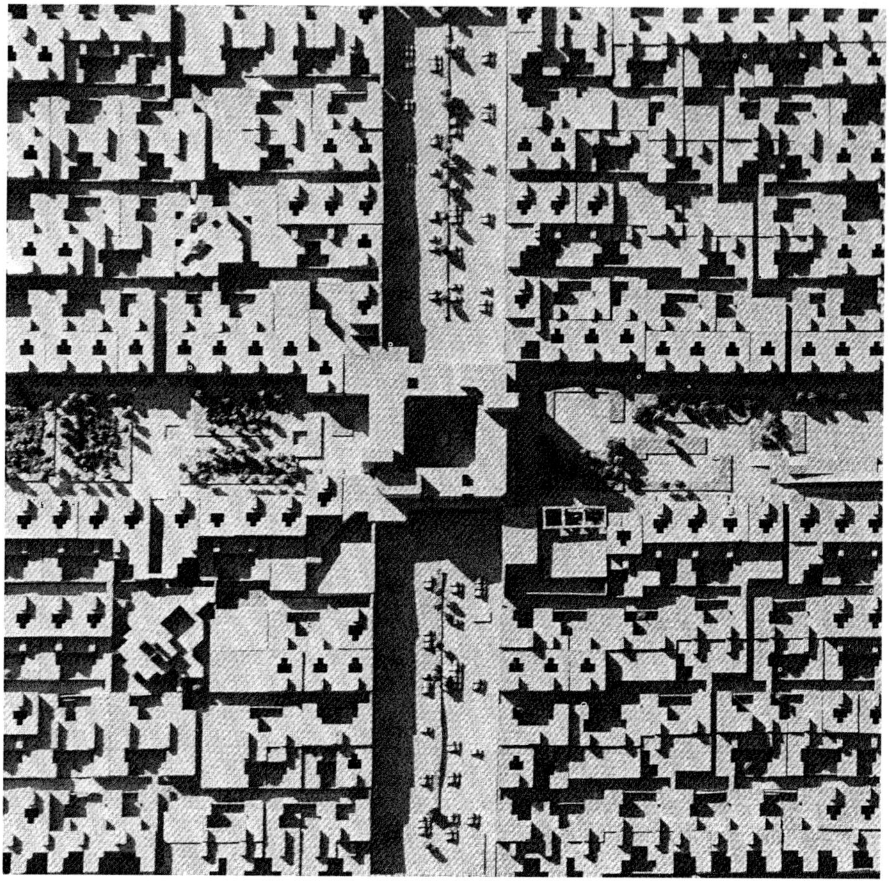

Fig. 10.2 Central residential quarter and neighbourhood plaza. Reproduced with permission of Kamran Diba

passages for the public (Diba 1980b) (Fig. 10.3). In individual buildings, thick walls and small windows on the shady side of the dwellings reduce the convection effect. Access to dwellings is normally through a porchway which provides a cool place for socialising. Roofs can be accessed by steps for the evening sleep, and brick screens provide shading without spoiling interior privacy (Diba 1986).

For Diba, the high-rise building is a serious enemy of a humane community. He believes that the cultural particularity of Muslim societies resists this kind of unfamiliar dwelling: 'We Muslims should avoid high-rises and create density horizontally' (Diba 2002: 121). Based on this argument, the majority of dwellings—about 80 %—are one- or two-storey houses; the rest are apartments. With regard to the housing concept, contrary to the Western notion of the house as an agglomeration of different rooms with particular functions such as living room, dining room and bedroom, the traditional concept of the room as a flexible unit was the departure

Fig. 10.3 Central shopping centre with deep arcades (Author 2011)

point, whereby large spaces are multi-purpose and potentially divisible. 'We planned two and three-room housing units which could become a four, five or six-room house as the family's standard of living improved' (Diba 1980b: 42). This flexible configuration, which Diba calls 'soft furniture' (Javaherian and Diba 2005), enables inhabitants to move between different spaces to avoid or enjoy the sunshine at different seasons. Rooms are medium sized at 5 × 5, 4 × 4, and the smallest at 3 × 4 m. On this flexibility Diba writes: 'Our goal was akin to writing a script for human interaction, anticipating all possible action and yet leaving room for spontaneous improvisation within the given architectural spaces' (Diba 1980b: 43).

In conceptualising the dwelling units, the courtyard was the main source of inspiration. Rooms are gathered around a courtyard, as the open space at the heart of the dwelling unit. This archetypal configuration, rooted in the Middle Eastern style of settlement and accommodation, has always been reinterpreted in poetic ways in Diba's architecture. In the Tehran Museum of Contemporary Art (1967–76) this archetype is formulated as an inner outdoor sculpture court, around which galleries are organised along a beltway. This feature is later used in the Administrative Building and Mosque of Jondi-Shapour University (1968–72), as well as in the Cultural Centre of Niavaran Garden (1970–78).

In Shushtar-e-No, a mid-technology approach has been employed using local technology in terms of materials and construction methods. Footings are of concrete, and the roofs are framed with steel beams supported by walls or engaged piers. The spaces between piers are filled by traditional brick barrel-vaults, over which concrete paving blocks are laid to form the final roof. Floor finishes are terrazzo tiles laid on concrete slabs. Most of the materials were produced locally. Not only was the contractor local, but the labour force consisted of unskilled labourers of whom 90 % were Iranian, and 10 % Afghan. The professionals employed on the project, including architects, engineers, and consultants, were 95 % Iranian (Diba 1986). Thus, this project was essentially dependent on local technology and a local work force, offering job opportunities for local artisans during the construction period.

A crucial problem with the Muslim traditional environment, according to Diba, is the uncontrolled nature of vehicular traffic circulation, the result of which is the attenuation of the quality and scale of the environment, and a reduction of the community as it becomes subservient to the car (Diba 2002). In opposition to this, Diba advocates the separation of the vehicular transport system from pedestrians. In Shushtar-e-No all parking zones are located at particular strategic points, to segregate the automobile from the community life. Parking plots are subsidiary to urban life and social interaction, and most public places are car-free zones.

Greenery is limited to courtyards and inhabitants are encouraged to create private green spaces within their houses. The reason is high maintenance costs of greenery and the possible future lack of maintenance services, which could degrade exterior green spaces and spoil urban vistas. Interior green spaces are sited close to the exterior, to provide shade for passengers. The only greenery is a few rows of trees along the sidewalks of the exterior highway and the main 'spine'.

This thematic review highlights distinct approaches implemented in this work that make it unique, such as simultaneous attention given to the local and the universal, highlighting social aspects of community life, and giving the building a tectonic character. These characteristics are quite enough to have convinced Kenneth Frampton, the architectural critic and theorist, to refer to this work as a prominent example of critical regionalism at the urban scale (Khan 2000).

10.4 From Utopia to Dystopia

Despite all these potentialities and capacities, the destiny of this project is one of an unfortunate decline, from an image of utopia to a state of dystopia. Utopias are sites of perfection. As the word 'utopia' suggests, they are a negation of place, a non-place. This non-placeness, however, indicates that utopia is essentially different from our common imagination of the 'place'. It illustrates an 'image' of perfection, to provide some pointers to a better condition. Shushtar-e-No has this kind of character.

For Diba, Shushtar-e-No was designed to construct a utopian image of a better society, detached from the problematics of the 'real, existing space', and thus a promise of a radical shift from the dominant paradigm of 'dwelling' and 'community' which had little to do with the authenticity and factuality of the given place.

The 'real space' of the dominant paradigm was that overwhelming semi-modernist approach in architecture and urban planning, which he resisted intensively. His utopian community was not the traditional homogeneous community of the past, nor the heterogeneous community of the present, but an ideal community of the future which mingles the past with the present in a constructive way. All the characteristics Diba intends to realise in his utopian world are either not-yet-extant, or what should exist. In this sense, his objective was designing an ideal place with a 'utopian vision', a place which is far removed from the unpleasant, overwhelming regularities of conformity, and opens new perspectives into the question of 'dwelling' and 'community'. This utopian vision, thus, was to be achieved through an 'architecture of resistance' which neither compromises the wishes of the client and policy makers, nor condones the rules of universal marketing. On the utopian vision of the project we read: 'Our goal was akin to writing a script for human interaction, anticipating all possible action yet leaving room for spontaneous improvisation within the given architectural spaces' (Diba 1980b: 43).

This utopian vision is observable in different aspects of the project, from its unique re-interpretation of traditional prototypes, to its encouragement of social interaction and enhancements to the public sphere, to environmentally progressive ideas of restricting private traffic. This is more understandable when the general trend of contemporary construction in Iran is taken into account, in which the intention was to follow universal patterns of mass housing and high-rise construction. However, this utopia was something 'imagined' in the mind, or an 'image' on the chapter. The realised place was soon transformed into a state of dystopia far from the utopian vision of its original conception.

At the beginning, the management of Shushtar-e-No was under the supervision of the Karoun Agro-industry Society. Soon after the first inhabitants were settled, major problems arose. The maintenance of the city was problematic, the ditches were dirty and drinking water was rationed. Immediately after the outbreak of the Iran-Iraq war, this area was overpopulated by war-stricken individuals, inhabitants of neighbouring villages, as well as multiple worker families living in the same house in overcrowded conditions.

Later, the degradation process did not stop, but was exacerbated. Houses were extensively adapted, without any regard for the context. Newly installed gas pipes were exposed on facades. For security reasons, metal fences were erected to protect low-level windows as well as entry points. Windows at a low height facing streets were frequently walled up. Air conditioning fittings were installed over some windows. Brick pavements were replaced by tarmac or whatever material found.

The lack of urban services is disturbing: everywhere is littered with rubbish and waste. There is no efficient sewage system. Public squares have been left unattended: they are abandoned terrains with no service and maintenance. Green space is scarce, with no regular irrigation (Fig. 10.4). Semi-private exterior courtyards, probably designed on the model of front yards of American suburbia (Javaherian and Diba 2005) and supposed to be mediatory green spaces between the private and public realms, are either devastated, or walled in. The shopping centre is dilapidated, far from its original state. The architect's intention in establishing a live city

Fig. 10.4 Public space left unattended (Author 2011)

centre has failed. Balconies (Ivans) have been extensively added to the interior space using any available material. More recently, most units were sold off by the municipality to public functionaries. The first phase which was planned for 4000 inhabitants has now more than 10,000 inhabitants. Consequently, the property values are very low compared to surrounding areas. In general, there is no vitality, dynamism, and sense of life in the town, and no sign of future improvements.

10.5 Discussion

The existing conditions of the complex can be observed from two perspectives: the initial gap between the utopian ideas of the designer and the real life of inhabitants, and the subsequent challenges that arose due to unexpected socio-political conditions.

Although this complex was never occupied by the people of the Kesht-o Sanat-e Karoun, the way current inhabitants have reacted to some of its architectural concepts, and recent architectural interventions, show that its current appearance would be partially similar even if the complex were inhabited according to the original agenda. There is enough evidence to claim that even the inhabitants originally envisaged for the scheme would react similarly to some of its

Fig. 10.5 Front gardens are mainly walled off (Author 2011)

architectural ideas, as similar treatments are observable in most low-income as well as middle-class housing projects all over the country. For example, exterior windows located at a visible height are inimical to residents' desire for privacy and security, so they have fixed this by haphazard walling and fencing off of windows. Most residents have also walled off semi-private front gardens to add to their sense of ownership, as they do not feel themselves responsible to the public (Fig. 10.5). A similar customisation has taken place with most balconies, which have been absorbed into the dwelling's interior for privacy or ownership reasons (Fig. 10.6). Moreover, cars are taken into the dwelling area by building into the courtyard of a garage-like space with wide doors, or are parked in front of them, and the public parking places are left unused. All these adaptations indicate that some architectural ideas were questionable from the beginning, regardless of their inhabitants, and in contrast to common understanding of the Iranian mode of dwelling and living.

On the other hand, a considerable part of the problem has originated from later unexpected conditions. The construction phase was stopped after the political unrest of 1979 and the subsequent revolutionary situation. First, the main objective of a community which could accommodate the white and blue collar workers of the Kesht-o Sanat-e Karoun Company was not achieved, as the company left the plan semi-realised after the Islamic Revolution. Later, war-stricken refugees were settled there, establishing an instance of what we can call, following Foucault, a 'crisis community' who are 'in relation to society and to the human environment in which

Fig. 10.6 Balconies merged into the interior (Author 2011)

they live, in a state of crisis' (Foucault 1998). War-afflicted people were the people in crisis, having been dislodged from their own homes due to the violence, and mainly consisted of poor people from the villages or poor cities of the region. The main idea of re-settling them was to provide them with a safe 'shelter', while the quality of life was less important. After the war, a proportion of the inhabitants left to return to their original homes, and their houses were occupied in an unsystematic way by new dwellers from the surrounding villages who had migrated to the urban areas. The state of crisis continued or even deteriorated, and this accelerated due to the inattention of the municipality and other governmental institutions in providing minimum urban facilities and public services.

Moreover, the initial idea of connecting Shushtar-e-No to the old city of Shushtar failed, due to its incomplete construction. Located outside the historic city, it established itself as an isolated realm tending to severe social and cultural segregation and disconnection. This was exacerbated by the establishment of what we have called a 'crisis community', a second hand community displaced from their origins and home. The result was a 'microcosm' characterised by its disadvantages, social problems, and lower-class migrants. It was a 'forgotten land'. This microcosm was identified with poverty, misery, and misfortune.

10.6 Conclusion

The Shushtar-e-No project is a distinctive work from a number of points of view. It is a housing project that provides residential facilities for labourers employed by an agro-industrial complex which has been developed at a period of economic boom. However, it tries to establish a unique community which goes beyond the dominant rules and principles of contemporary construction, offering a third way in which tradition and modernism are effectively linked together. It thus promises a radical shift from the dominant architectural and building paradigm essentially influenced by modernism and the International Style. The initial vision is to some degree utopian, in the sense that an ideal, but inexistent manner of 'dwelling' and 'community' is projected. But this 'image,' which seemed to work perfectly at a conceptual level, was never realised on the ground; the result was a dystopia per se, far from the project's initial objectives and ideals.

This dystopia was resulted from two main impediments: firstly, the gap between the designer's imagined ideal and the user's actual handling, as indicated by the later treatment by the inhabitants; and secondly and mainly, the subsequent unexpected socio-political conditions, which led to the misuse of the original project for short-term problem solving. Currently, Shushtar-e-No suffers from a plethora of serious problems and challenges.

Shushtar-e-No oscillates between two senses of heterochrony: on the one hand, the nostalgic time of the traditional architecture and city to which the plan of the city is indebted; and on the other, the solid time of the recent past, namely the disturbing memory of the war and its terrible aftermath. These two sides of heterochrony point to a disturbing memory from which individuals prefer to escape. While this community is open to the public, it is unattractive to visitors due to its catastrophic conditions. Nevertheless, visitors could potentially be attracted by the distinctive approach that its famous architect employed in creating its design. It is visited by architects to learn from its hidden lessons, or by sociologists to study its social challenges: not as a solution, but as a problem.

Shushtar-e-No is a heterotopia *par excellence.*

References

Abrahamian E (1982) Iran between two revolutions. Princeton University Press, New Jersey, p 140

Abrahamian E (2008) A histoy of modern Iran. Cambridge University Press, New York

Diba K (1980a) The recent housing boom in Iran-lessons to remember. In: Safran L (eds) Housing: process and physical form, Philadelphia, Aga Khan Award for Architecture, pp 38–40

Diba K (1980b) A case study: design concept of Shushtar New Town. in Linda S (eds) Housing: process and physical form, Philadelphia, Aga Khan Award for Architecture, pp 41–44

Diba K (1981) Buildings and projects. Hatje Publisher, Stuttgart

Diba K (1985) Shushtar New Town, Iran. MIMAR 17:49–53

Diba D (1986) Shushtar New Town, technical review summary, Aga Khan Award for Architecture [online]. Available from: http://archnet.org/publications/178. Accessed May 2012
Diba K (1998) Shahrake Shushtar dar Nemaieshgahe Jahani. [Shushtar New Town in International Exhibition], Me'mar 1, Tehran, p 87
Diba K (2002) What Islamic Architecture is not. In: Petruccioli A, Pirari K (eds) Understanding Islamic architecture. Routledge Curzon, New York, pp 119–124
Diba K (2010) 'Baghi Miane do Khiaban' [A garden between two streets], in converstation with Reza Daneshvar. Alborz Publisher, Paris
Diba D, Beheshti M (2004) Trends in Modern Iranian Architecture. In: Jodidio Philip (ed) Iran: architecture for changing societies. Umberto Allemandi & C, Torino, pp 31–41
Foucault M (1998) Of other spaces [online]. Available from: http://foucault.info/documents/heteroTopia/foucault.heteroTopia.en.html. Accessed June 2012
Habibi Seyyed M (2008) As Shar ta Shahr [From Shar to City]. Tehran, Daneshgahe Tehran
Habibi Seyyed Mo, Hourcade B (2005) Atlas of Tehran metropolis. Pardaz va Barnamerisie Shahri Publications, Tehran
Javaherian F, Diba K (2005) Shushtar New Town. InIraj.E et al. (eds.) New towns, Tehran, new towns development sooperation, pp 15–36
Kamali M (1998) Revolutionary Iran. Civil Society and State in the Modernisation Process, Brookfield, Ashgate
Khan H-U (2000) Expressing identities thorough architecture, from colonialism to pluralism, In: Frampton K (ed) World Architecture 1900–200: a Critical Mosaic, Beijing, China Architecture and Building Press, pp. xx–xxxv
Madanipour A (1998) Tehran, the making of a metropolis. Wiley, Chichester
Madanipour A (2006) Urban planning and development in Tehran. Cities 23(6):433–438
Miraca L (1996) Contemporary Iranian architecture in search for a new identity. Environ Des J Islam Environ Des Res Centre 1:52–91
Mirmiran Seyyed H (2004) Public buildings in Iran: 1920 to the present. In: Jodidio P (ed) Iran: architecture for changing societies. Umberto Allemandi & C, Torino, pp 39–41

Chapter 11
Tehran and the Lost Nature

Parisa Mirsadeghi

Abstract Located at the foot of the Alborz Mountains, Tehran once had a reputation for its many gardens, trees, and natural beauty. It was because of this rich natural asset that the city was even dubbed '*Chenarestan*', or the land of *Chenar* trees. Today, there is not much left of this natural element due to decades of rapid urban development and expansion. This chapter presents an analytical account of how this asset came under such heavy threat that there is now a clear scarcity of green spaces in the city. The chapter first traces how social and economic factors, along with the process of modernisation have, over years, contributed to this destruction. An analysis follows of the cultural factors playing parts in this process. The relationship between governmental, public, private factors and the role of civic society as a regulating body are used as a framework for analysis. In the final section, a case study of Darband neighbourhood is offered to illustrate the points made. The destruction of Darband's rich natural environment exemplifies the wider process of incongruent development and its consequences for Tehran. The case study demonstrates how various factors discussed in the chapter operate to shape the concrete reality of the city.

Keywords Urban development · Green spaces · Destruction of nature · Tehran · Darband

11.1 Introduction

The Urban Revolution takes its toll on natural territories upon its arrival. As processes of urbanisation and industrialisation intensify, natural environment, wildlife, and the intact nature inevitably undergo profound transformations. Unfortunately, agricultural and natural resources are targets for urban construction. As a result,

P. Mirsadeghi (✉)
Shahid Beheshti University, Tehran, Iran
e-mail: Parisa.mirsadeghi@gmail.com

F.F. Arefian and S.H.I. Moeini (eds.), *Urban Change in Iran*,
The Urban Book Series, DOI 10.1007/978-3-319-26115-7_11

these sites are prone to gradual destruction and replacement with roads, buildings, and urban installations (Mumford 1956).

The constant expansion of urban centres not only absorbs the peripheries, but also allures the population of the remaining non-urban areas to the cities by offering them various professional, recreational, cultural and service-related opportunities.

Nevertheless, and due partly to the people's tendency towards nature, and partly in reaction to various irritations and annoyances of the urban lifestyle, e.g. over-population, as well as over-systematised and impersonal city life, many of them renounce this kind of life and look for possible getaways. The attractions of the city, however, and people's dependency on cities for work and socialisation prevents them from breaking their ties to it altogether, making them return to the city every time they leave. Consequently, suburban settlements are created, which in turn need more roads, urban infrastructures, and public utilities, augmenting the environmental pressures on the ecosystem (ibid.).

Tehran exemplifies this with its consecutive profound changes and transformations and expansion during its relative short time of being the country's capital (around 230 years). This is how an idyllic village in the vicinity of the city of Ray (Shahr-e Ray) has metamorphosed into the gigantic megacity of today in relation to which Shahr-e Ray is but a peripheral outgrowth. The continuous and intense process of development and construction in Tehran has rendered the city an ever-changing entity.

These processes of construction has led to extensive changes and transformations in Tehran's natural terrains and surrounding rural areas. Tehran was once famous for its numerous Oriental Plane trees (*Platanus orientalis*). Travellers who passed through Tehran used to find the city covered with parklands and fruit orchards (Madanipour 1998). The city, however, has lost many of these orchards and their trees due to its rapid growth and development.

Urban green spaces are crucial for absorbing and neutralising pollutants, production of Oxygen, and counteracting environmental threats in a city like Tehran (Parivar et al. 2008). The function of parks and green spaces in Tehran, however, goes beyond merely environmental. In fact, not only does the scarcity of green spaces in Tehran pose serious environmental problems, it also contributes to the city's failure to meet the nonmaterial and psychological needs of its inhabitants (Soltani 1986).

Public places, city squares and plazas, which offer citizens the chance to 'pause', are virtually non-existent in Tehran. This is a city in which everyone is always on the move; the streets are only paths for getting from one place to another and people just move through them hurriedly. Consequently, city parks and green spaces are practically the only available spaces in which it is possible for people to take a moment to rest and refresh.

Unfortunately however, these green spaces are merely isolated spots of greenery scattered here and there without any consistency or connection to one another. They do not comprise the background settings that connect human constructions and buildings. On the contrary, it is always the urban landscape that surrounds the green stains on the map, shutting these green spaces off from each other and from people.

Casting a glance at Tehran's formation and development, especially in the recent decades, this chapter aims to explore the major influences in the changes and transformations occurred to the natural environments of the city. It also aims to look into the city's cultural structure which distinguishes its changes and transformations from those in other megacities, as well as the ways in which these changes have taken place in Tehran. Finally, the Darband region will be presented as an example demonstrating the issues discussed.

11.2 Contributors in the Destruction of Tehran's Nature

As with many other megacities, issues such as population growth, rising land prices and transport significantly contribute to the changes and transformations Tehran has witnessed, and thereby in the destruction its nature has suffered. A brief survey of Tehran's growth and development, particularly within the last century, manifests how these three factors have come to shape the Tehran of today and caused the destruction of its nature.

11.2.1 Demographic Developments

The growth of population and the influx of immigrants into Tehran have put considerable pressure on the city's natural environment. Although the irregular development of the city, the rapid population growth, and the constant arrival of immigrants have intensified in the past few decades, they have, in fact, been at work since Tehran's selection as the capital city.

Around World War II, the city did not experience considerable growth, even though the population was fast increasing. After the war the urban expansion speeded up resulting in many villages and Qajar gardens being demolished to accommodate the construction and development of urban facilities. The same happened to vast areas of agricultural lands in this phase of urbanisation (Mehdizadeh 2003).

In the post-war decade, as the city expanded, it engulfed industrial centres and facilities previously located outside city boundaries: a development which in turn added to environmental problems of Tehran. Simultaneously, the advances in construction technology and equipment further accelerated the pace by which rural lands were being encroached by the city.

Between the 1950s and the 1980s, the population of Tehran quadrupled, an increase which could mostly be attributed to the continual process of village to city migration triggered by various political and economic political motivations (Tehran's Ecological Study Report 1988).

In the 70s, the increased oil income and the resulting monetary expansion, most clearly manifest in Tehran, added momentum to the processes of immigration and urban constructions. These have been constantly on the rise ever since (Tehran's Ecological Study Report 1988).

11.2.2 The Increase in Land Prices

As a result of constant population growth and immigration, Tehran has been grappling with a shortage of space. These, together with economic policies in which urban development was encouraged as a means to stimulate economic growth rather than serve urban needs, resulted in soaring land prices. With this increase, old gardens and parks were destroyed and replaced with high density and profitable residential buildings. Although cutting trees was forbidden by statute within the jurisdiction of Tehran municipality, estate owners were given permission to cut down sick and old trees. This created a legal loophole which could be exploited by owners and constructors to rid themselves of trees (by intentionally leaving them to wither) whenever the land could be used for more lucrative ventures (Madanipour 1998).

What is more, due to expensive land prices and high costs of living in Tehran, immigrants were usually pushed to the margins of the capital, which in turn pushed forward the borders of the city and led to further destruction of natural resources and absorption of more lands into the city.

As mentioned in Tehran's Ecological Study Report (1988), land prices in Tehran are among the highest for urban land globally. This has created a booming market for land speculation, impeding successful design and execution of long-term urban plans aimed at reforming urban planning and administration procedures. In order to enhance the physical structure of Tehran, it is necessary to oversee the allocation of spaces in view of obtaining a sense of spatial balance in which the land-use hierarchies are observed. In the face of soaring land prices and the absence of such spatial balance, not much land is available for creation of green spaces.

11.2.3 Transport

The process of modernisation and the introduction of motor vehicles drastically altered the appearance of Tehran. The construction of new streets and widening of existing roads to make them suitable for cars brought about a new era in urban development. As Tehran's development was quite incongruent, expansion of the road system overshadowed other areas such as the sustenance and provision of green spaces.

Population growth, increased income and thereby purchasing power, changes in lifestyle, as well as excessive production and import of cars are among reasons for the construction of more streets, roads, and highways (Tehran's Ecological Study Report 1988). This ever-perpetuating need for more roads is believed to be met through demolition of residential areas and buildings to make room for new planned roads and construction of double-deck highways to tackle the problem of spatial scarcity.

Meanwhile, urban green spaces and parks have fallen prey to profit-oriented road construction projects. The prioritisation of road construction over preservation of green urban spaces, as well as the absence of legal protections against demolitions are the two factors still threatening the natural capital of Tehran. Consequently, when solutions are sought for the shortcomings of the city's transport system, available options are usually single-minded, just focusing on the ways in which heavy traffic can be managed with little concern about natural environment.

This problem is manifest in the policy adopted during the construction of Tehran's subway system which, thanks to its economic use of land and less polluting nature, is considered as one of the most environmentally friendly means of public transport. In this case, local neighbourhood parks were the first places sacrificed to solve the problem of space scarcity. This was exemplified by demolition of two such parks to build Gheytariyeh and Sadr metro stations.

These destructive trends are still in existence today, leaving no harmony and balance between green spaces and other urban areas in Tehran and casting a soulless shadow over the remnants of green spaces.

11.2.4 Culturally Destructive Factors

Apart from the factors mentioned above, which can be regarded as inevitable repercussions of the constant urban growth and development in metropolises, natural environments of Tehran came under threat from two main processes of intrusion and the absence of controls restraining these intrusive forces. The first threat is the intrusion of state and government forces into the public sphere and the second stems from the intrusion of the private sphere into the public sphere. The weakness of civic society, as the force controlling the relationship between these spheres is yet another factor which works against the preservation of Tehran's natural capital. Below are explanations of each threat:

- The modes of civic administration exercised in Tehran display a stark dearth of rationality, so that regardless of who shoulders responsibilities, the system is incapable of protecting the natural ecology of the city. The social structure of today's Iranian society and the socio-cultural scene of Tehran for that matter indicate a strong tendency toward the short term (Katouzian 2004). In this sense, social arrangements and institutions are constantly subject to rapid changes as a result of which the previously established goals, values and modes of practice are abandoned, losing their executive power and being replaced by new ones. More specifically speaking, Tehran municipality has been a very politically sensitive institution in the past few decades and has, therefore, undergone many politically influenced changes: every managerial replacement has resulted in fundamental changes in policies.

  A key concept in understanding these processes of constant change is *haste*. Kazemi (2012b) has explained the role this concept plays in shaping the contemporary everyday life of Iranians by expounding what he calls 'the culture of hurriedness.' According to this frame of reference, citizens of Tehran are constantly in a state of hurry. The same condition applies to the authorities; in the state of *haste*, the urgency and importance of every single project or policy decision is magnified to the extent that putting them in context with other projects and adjusting them accordingly is often paid little attention. The destruction of several local parks to make place for the construction of metro stations, e.g. Gheytariyeh and Sadr stations, as well as the Sadr highway-side trees, uprooted in the process of building the double-deck freeway, can both be considered as lucid examples of this lack of reason in planning and prioritising.

- In the Iranian society, the invasion and intrusion of one sphere of life by the other does not happen in a linear, top-down manner: alongside the top-down forces of the government which invade the public and private sphere, there exists the intrusion of the private into the public sphere in various forms (Kazemi 2012a). Motivated by immediate private interests and against the public good, citizens lay abusive hands on public spaces and flout the law using various justifications. Even though this conflict of interests is not confined to the Iranian society, it is the systematic nature and the magnitude of such violations which renders the situation problematic in this country. In this sense, the legitimacy of the social contract is jeopardised when breaching the law is no longer an exception, but the prevailing modus operandi. Behaviours demonstrating such mentality include, among others, intentionally drying tree orchards in order to get permission for high-rise constructions not to mention the illegal changes of use of agricultural lands, none of which limited to the capital.
- In view of the interplay between these opposing forces, a strong civic society could have played a significant role in aligning forces and counteracting their destructive effects (Kazemi 2012a). The presence and activities of powerful non-governmental organisations in many megacities have proved extremely valuable for the preservation of the environment. Unfortunately, however, such organisations in Iran are quite weak and, historically speaking, a civic society in the above sense has never existed in Iran. As a result, the absence of watchdog institutions and social control mechanisms which could create some balance, suggest that the issue of proper management of natural spaces will remain a challenge for the capital.

11.3 Patterns of Change in Tehran's Natural Environments

Regardless of the reasons behind the changes, it is important to investigate exactly how this process of transformation has occurred. These changes and their consequent destructive impacts on the city's natural resources can be laid out in two different discernible patterns: internal and external.

11.3.1 The Internal Pattern

In his article about Tehran, Rezaee Rad (2011) likens the city to a hyperactive organism which needs much more oxygen than is now available, resulting in a full-scale crisis. Should we break down the metaphor to its components, what Tehran so direly needs is 'space'.

Tehran is confronted with the problem of spatial scarcity. Existing spaces are perpetually devoured so that those which could still serve legitimate needs, are short-sightedly replaced to make room for other perceived needs. For example, two or three-storey houses are replaced with high-rise buildings and those labelled as 'pick-axe' buildings[1] (*Sakhteman-e-Kolangi*) are deemed suitable for demolition to build new ones. Green spaces are not treated differently. In spite of the existence of protective laws, the city has devoured green spaces and converted them to roads and buildings in the process of its growth and expansion. The laws in operation still have their own shortcomings. Demolition of small gardens and green spaces (under 500 m^2), for instance, does not require any special permit from the authorities (Madanipour 1998). Developments in Farahzad neighbourhood, where a great number of once lavish tree orchards were replaced with high-rise structures over a decade, is another case in point illustrating the inadequacy of legal protection of green spaces.

11.3.2 The External Pattern

The internal spaces do not satiate the hyperactive organism. The city, therefore, needs to devour external spaces in order to sate its appetite for endless expansion. The consequent urban sprawl perpetuates the rapid growth for two main reasons. Firstly, the high prices of land and cost of living in Tehran forces those who cannot afford such expenses but are nonetheless attached to the capital by professional ties to move to nearby suburbs. Secondly, the relatively more pleasant weather and environment of some suburban areas such as Lavasan and Kordan (in Karaj), have proved appealing (Tehran's Ecological Study Report 1988).

Various attractive features of these suburbs have raised land prices in these areas to levels occasionally higher than those of many areas in Tehran. This in turn raises similar concerns to those of Tehran. This continuous process of expansion towards more intact peripheries of Tehran leads to a lucrative real estate market in suburban areas previously not having experienced such levels of activity and change. Like in many other megacities, this process of rapid development and construction in the

---

[1]An expression referring to buildings more than 30 years old which in the Iranian context are considered to have passed their useful life and therefore are to be demolished. This term was used by Homa Katouzian in "The Short-term Society: A Study in the Problems of Long-term Political and Economic Development in Iran".

peripheral areas of Tehran generates an increasing demand for more roads, cars, infrastructural facilities, and service centres which all have negative effects on the natural environment. The transformation of Darband neighbourhood exemplifies this point. The arrival of wealthy population who sought a refuge from the troubles of the megacity brought about a new wave of construction which took its tolls on the natural environment of Darband.

11.4 Darband as a Case Study

In order to clarify the issues discussed thus far, we examine a case by drawing a broad comparison between the old and the new Darband. Darband is located in northern Tehran and has a reputation as one of the most pleasant natural environments in Tehran. The neighbourhood is currently comprised of two different urban fabrics. The main reason for choosing Darband as a case study is the significant difference between these two parts. On the one hand, there is the old Darband with its organic fabric. On the other, new mass inharmonious constructions are giving shape to a different fabric of high-rise buildings rapidly devouring the natural environment of Darband.

Darband has more than two hundred years of history (Niavarani 1950). The pleasant weather and the idyllic nature played important roles in the early formation of the region in which most residents were farmers (Sanad-e hoviat-e mahaleh 2007). As Tehran expanded over time, Darband turned into one of the nearest summer resorts for the residents of Tehran. The wealthier families were attracted to the region because of its weather and nature. The Sa'd-Abad palaces were also built nearby for the royal family to enjoy these natural features. The construction of guesthouses, administrative offices, a police station, and power plants followed (Sotoudeh 1993).

Population growth in the past decades has turned Darband into a well-defined and integrated neighbourhood of the city so that even though Darband still remains as an attractive tourist resort, considerable numbers of people also live there permanently. The development of Tehran has brought about remarkable changes around Darband River in the past two decades. These changes have resulted in the rapid erosion of the area's historic fabric. What follows is a closer look at the two different fabrics of today's Darband (Fig. 11.1).

11.5 The Old Darband

The old fabric of Darband stretches from the edge of Darband Street up to the foothills of the mountains. The main characteristics of this quarter have been shaped under the strong influence of its surrounding environment. The organic fabric of this part, formed in harmony with the topographic features of the land, has helped

Fig. 11.1 A view of Darband form North (Author 2012)

develop a strong integration between the urban fabric and nature. The buildings placed amid green spaces or courtyards are not in conflict with natural elements of the area. In contrast to monumental modern apartment blocks, these buildings have a human scale.

In this part of Darband, streets are usually narrow and there are many stairways which mark the steep landscape, a clear sign that the urban fabric was developed prior to the introduction of cars. A defining characteristic of this quarter, this fabric shows a remarkable level of compatibility and congruence with natural elements, although it undeniably causes significant hardship for the residents, especially during the cold seasons of the year (Fig. 11.2).

Due to increased land prices and the strong attraction that Darband has had for many construction contractors in the recent years, this old quarter has become ever more prone to lose its organic fabric and its intact nature as old houses are demolished, trees are cut down, and new streets are built to make room for cars.

11.6 The New Darband

Behind the houses of the old Darband, climbing up the stairways, one encounters a scenery of wide streets and high-rise buildings in place of what used to be orchards and garden houses at the foot hills of the mountain. This sudden drastic change in

Fig. 11.2 The old fabric of Darband (Author 2012)

the urban fabric within such proximity indicates a fundamental shift in the logics of the built environment. While the organic development of old Darband was dominated by the concept of interaction and preservation of nature, as the new scenery spreads out the emphasis on harmony with nature is replaced with a profit-oriented incongruent urban development logic (Figs. 11.3 and 11.4).

The main consequence of this shift is that priorities of this process of urban development seem to serve interests other than what the neighbourhood requires for its organic development over time. In the new urban fabric, priority has been given to streets, cars, and constructing profitable high-rise buildings. Preservation of the region's natural capital is not taken seriously if not ignored altogether. Providing a spectacular view or perfect natural lighting for the newly bulit apartments becomes a priority even if this would entail compromising these very qualities for others (Fig. 11.5).

Alien to the particularities and features of the topography of the area it is built on, the urban fabric of this part of Darband has been homogenised by the standards and logics of the real estate market which has given shape to it. This has a detrimental effect on the natural element of the area which used to be the most

Fig. 11.3 The new fabric of Darband (Author 2012)

Fig. 11.4 The new fabric of Darband with wide streets (Author 2012)

FIg. 11.5 High rise buildings which obstruct the view of the mountain (Author 2012)

distinctive feature of Darband. This is to the extent that not much difference can be discerned between the fabric of the new Darband and those of any other part of Tehran.

11.7 The Role of Cultural Factors in the Transformation of Darband

The needs of the new residents of Darband, such as the need for streets to allow cars to pass through, should not be dismissed when investigating the process of transformation this neighbourhood has gone through. The extent of transformation, nevertheless, goes far beyond mere provision for these needs.

In blatant disregard for nature and the local ecosystem, the excessive construction permissions, issued in disregard for the particular requirements of the area, are gravely problematic. It can be concluded that the logics of the market and profiteering have in this case has been the sole priority and frame of reference for decision-making, resulting in a complete disregard of other vital civic and environmental considerations.

What is more, with weak civic institutions in place, individuals invade the public sphere and exploit its resources. As mentioned, there are various ways for mass construction contractors and private owners to contravene construction rules and regulations. Often, rises in land prices in a particular region of the city are followed by intervals during which construction permissions are more easily granted and deviations from the law are handled more permissively and excused with lenient fines. In this state of affairs, civic institutions can shoulder the responsibility of intervening in support of the public good to protect the public sphere against the invasions of both government and the private sector. Sadly though, the relationship between local city councils and the citizenry are often fragile. The councils' programmes and activities are usually formed and implemented without the engagement of local people and consideration of their needs, and the connection between councils and local residents is often limited to election times. Under these circumstances, the absence of non-governmental organisations and civic institutions defending the public good, including the right to a balanced ecosystem is quite evident.

11.8 Conclusion

The natural environments of Tehran, once considered as one of its most valuable assets, have been destroyed in many ways and at a rapid pace: a trend that can be observed in Tehran as a whole as well as in regions of particularly valuable natural environments. These environments have often historically had a double function in Tehran: they are the core reason why people are attracted to these areas and also the first elements harmed as a result of people's arrival and their activities. There are either no controls to curb and prevent destruction, or in cases such controls are introduced, they have limited influence. The improvement in the NGO sector and private organisations as well as the empowerment of a civic society can play a significant role in raising the citizens' knowledge and information about their civil rights and duties.

References

ATEC Consultants (1988) Tehran's ecological study report: migration and spurt expansion. ATEC Publications, Tehran

Katouzian H (2004) The short-term society: a study in the problems of long-term political and economic development in Iran. Middle East Stud 40(1):1–22

Kazemi A (2012a) How 'the public' relates to governmental and private forces. Lecture to sociology of everyday life in Iran. Rokhdad-e Tazeh Research Institution, 12 June 2012

Kazemi A (2012b) Piramoun-e Farhang-e Ajaleh dar Jame' Irani (On the culture of hurriedness in the Iranian Society) (online). Available from http://akazemi.ir/fa/?p=467. Accessed 4 July 2012

Madanipour A (1998) Tehran: the making of a metropolis. Wiley, New York

Mehdizadeh J (2003) Shahr va Trikh (City and history). Joustarhaye Shahrsazi 1(3):16–26

Mumford L (1956) The natural history of urbanization. In: Thomas WL (ed) Man's role in changing the face of the earth. University of Chicago Press, Chicago, pp 382–398

Niavarani O (1950) Tarikhcheh-e Shemiranat (History of Shemiranat). National University Press, Tehran

Organisation of Tehran Beautification (2007) Sanad-e Hoviat-e Mahaleh (The neighborhood identity document). Office for Official Publication of the Organisation of Tehran Beautification, Tehran

Parivar P, Sotoudeh A, Yavari A (2008) The analysis of urban green spaces of Tehran. Mohit Shenasi 34(45):73–87

Rezaee Rad M (2011) Hyperactive organism. Art Q Mag 8(36):16–31

Soltani K (1986) Negahi bar Fazaye Sabz-e Tehran (A brief outlook on green spaces in Tehran). Mohit-e Zist 11(13):40–47

Sotoudeh M (1993) Joghrafiya-e Tarikhi-e Shemiran (Historical geography of Shemiran). Moaseseh-e Motale'at va Tahghighat, Tehran

Part IV
Shaping Iranian Cities (Zooming Out)

Chapter 12
The Relationship Between Urban Governance and Sustainable Urban Development in Iran

Naser Barakpou and Ramin M. Keivani

Abstract Cities in Iran face major challenges in all domains of sustainability relating to environmental, social and economic dimensions of development. Environmental challenges include high and ineffective energy consumption, destruction of natural environment, pollution of air, water, soil and other related problems. On the social and economic fronts, poverty and inequality in access to public services and infrastructures and employment has led to rapid growth of informal settlements and informal economic activities inside and on the fringe of medium and large cities. Previous research and reports on evaluating sustainability of Iranian cities and regions have largely focused on assessing sustainable development indicators in environmental, social and economic areas. As a result, they have neglected to examine underlying procedural issues affecting sustainability, e.g. urban governance, and the relationship between substantive and procedural aspects. The main aim of this chapter is to examine the institutional and social capacities for moving and progressing towards sustainable urban development in Iran. The chapter adopts a systematic review or meta-analysis approach and brings together results of a number of previous studies undertaken by the authors and other researchers. The findings indicate that there are not sufficient social and institutional capacities for progressing towards sustainable cities. Nevertheless, the chapter contributes to better understanding of the potential that already exists, the barriers that must be overcome and the role that main agents play in shaping and utilising these capacities.

Keywords Sustainable urban development · Urban governance · Social capital · Institutional capital · Iran

---

N. Barakpou (✉)
University of Art, Tehran, Iran
e-mail: barakpoo@art.ac.ir

R.M. Keivani
Oxford Institute for Sustainable Development, Department of Real Estate and Construction, Oxford Brookes University, Oxford, UK
e-mail: rkeivani@brookes.ac.uk

F.F. Arefian and S.H.I. Moeini (eds.), *Urban Change in Iran*,
The Urban Book Series, DOI 10.1007/978-3-319-26115-7_12

12.1 Introduction

The idea of sustainability has been part of the discourse of development at least since 1987 when the Brundtland report was published. Yet we are still confronting vast challenges that are to different degrees present in the cities of both global North and South (Keivani 2009). However, especially due to accelerating urban trends in the South, combined with serious infrastructural and institutional deficiencies, the concern with unsustainability of development, particularly in cities, and neutralising its negative effects is far greater and more urgent. Iran is a case in point where in the past 50 years the urban population has increased from 30 % to over 70 % (Statistical Centre of Iran 2012) and at the same time the country is facing major sustainability challenges in both rural and urban areas.

These challenges are manifest in all domains of sustainability and at different spatial scales from national to local neighbourhood levels. Environmental challenges include high and ineffective energy consumption, destruction of natural environment, pollution of air, water, soil, sound and other related problems. For example, World Bank (2005) shows that in 2002, the annual cost of environmental degradation in Iran (including water, land and forest, air, waste, coastal zone and CO_2 emissions) was estimated at 8.8 % of Iran's GDP (equivalent to US$10 billion). From this amount, US$8.4 billion is due to damages to the country's national economy and US$1.6 billion to the global environment through greenhouse gases (CO_2 etc.) emissions. A clear manifestation of this can be seen in the accelerated use of natural land resources leading to a 50 % reduction in the size of forests in the last 50 years and a degradation of water resources across the country both in quality and quantity (WGES 2009). Emission of pollutants and greenhouse gases emissions, such as SO_X, NO_X and CO due to Fossil fuels combustion has led to massive pollution of air and other vital resources, especially in big cities and metropolises (ibid).

Furthermore, the latest reports of the Iranian Ministry of Energy (2011) and International Energy Agency (2010, 2011) indicate that the intensity of final energy consumption in Iran is approximately twice the world average; or, in other words, Iran energy efficiency is almost half of the average world indicator. Generally speaking, different environmental indicators show increasing unsustainability of development in recent decades in Iran.

In the social domain, we can point to the rapid development of informal settlements and the population increase within such settlements in the last 20 years in the margins and inside the medium and large cities of the country as a significant indicator of development unsustainability. Accordingly at least 20 % of the urban populations of the country live in such settlements and the trend is increasing. For instance, the data released by Urban Development and Renovation Company, affiliated to the then Ministry of Housing and Urban Development, in 2006 reveals that in some medium cities of the country, about 1/2 of the urban population were living in informal settlements (MRUDI 2012). Zahedan with a population of over 500,000 and Sanandaj with a population of over 300,000 are among such cities.

The population of informal settlements in Mashhad (the second highest populated city in the country with 2.5 million inhabitants) has grown from about 100,000 people (16 %) in 1976 to over 900,000 (33 %) in 2007.

In the field of economy, the irregular financial resources of municipalities, and their severe dependence on unsustainable revenues gained through building development fees and levies are among major obstacles blocking the road to sustainable urban development. This was instituted in 1983 with the introduction of the financial self-sufficiency policy for municipalities as a means of coping with the reduction in central government funding allocation. In Tehran, the revenues gained through construction fees were among the most important items of unsustainable revenues for municipalities. These revenues include levies on buildings and lands, levies and duties on planning permissions, levies and duties on building permits and lands demarcation, levies and duties for omission of car parking, levies on terraces and balconies, revenues gained through adjacency rights, and revenues from the implementation of Article 100 of Municipalities' Law. Fluctuations in urban construction activities and any kind of recession or boost in the construction sector heavily affect such revenues and therefore, has an immediate bearing on municipalities' financial capacity. Based on different studies, during the 15-year period from 1992 to 2006, this revenue source has always accounted for over 75 % of municipality revenues and in some years (2000–2001), it has even reached as much as 90 % (Pourzarandi et al. 2007).

Previous research and reports on evaluating sustainability of Iranian cities and regions support these claims (e.g. WGES 2009; World Bank 2005). In fact such studies have reinforced sustainability considerations of policy documents and sustainability regulations on urban development in Iran in recent years. Yet the situation has progressively worsened! The reason for this may be that the approach so far has been largely focused on assessing and strengthening what Faludi (1973) has described as 'substantive' issues in planning, i.e. methods and application of planning such as identifying and measuring sustainable development indicators in environmental, social and economic areas. As a result, and again using Faludi's terminology, they have neglected to examine underlying 'procedural' issues affecting sustainability, e.g. urban governance, and the relationship between 'substantive and procedural' aspects. Indeed, recent international studies, such as the Discus Project (Evans et al. 2005, 2006) have indicated that procedural factors, including institutional and social capacities are fundamental for progressing and achieving sustainable development goals. Based on these premises, the main question of this research is: from a procedural viewpoint, are there sufficient social and institutional capacities for moving and progressing towards sustainable urban development in Iran?

In order to answer this question, Sect. 2 will discuss the relationship between urban sustainability and urban governance. In Sect. 3, following a brief discussion on methodology, the chapter examines the state of sustainability of Iranian cities. After that, and based on the theoretical framework resulted from literature review, the main procedural factors for progressing towards sustainability of Iranian cities are discussed in Sect. 4. The chapter finally concludes by presenting a discussion about what may be the most appropriate future agenda for research on the relationship between sustainability and governance in Iranian cities.

12.2 The Relationship Between Urban Governance and Sustainable Development

The relationship between urban governance and sustainable development may be studied from both theoretical and practical aspects. Theoretically, the role of governance in realisation of sustainable development is emphasised in different texts. Part 3 of Agenda 21 that includes Chapters 23–32 is devoted to the role of key, effective groups in promoting sustainable development goals. In the introduction to this part the necessity of participation and commitment of all social groups to effective implementation of the agreed objectives, policies and mechanisms in all fields stated in Agenda 21 is emphasised. Subsequently the role of women, children and the youth, the role of native inhabitants, cooperation with non-governmental groups, local authorities, trade unions, industry and business sectors, scientists, experts, and farmers are discussed in greater detail. In Chapter 28, the authorities are requested to pave the way for full participation of the women and the youth in decision-making, planning, and implementation processes and consult with citizens and social, industrial, and professional organisations on data gathering and obtaining consensus on sustainable development strategies. These themes are what can be called, in the broad sense of the word, the process of urban governance (Evans et al. 2005).

In addition to Agenda 21, the role of governance is highlighted in other texts on sustainable development (e.g. Rogers et al. 2008). The sum of different analyses presented in these texts is that the state, government, and formal institutions, do not have the required capacity, ability, knowledge, and resources for realisation of the objectives of sustainable development. Therefore, the participation of the local communities, cooperation of non-governmental organisations, tapping on the resources and capacities of different social groups, and generally, the process of urban governance are indispensable for realisation of sustainable development.

From a practical perspective, the relationship between urban governance and sustainable development can be considered from two aspects. The first is the analysis of previous experiences and actions to unveil the weaknesses and inabilities of the government in promoting the objectives of urban development. For instance, we can mention the emergence of citizen's participation strategy in response to government's inability to revitalise deteriorated urban areas in the 1950s and the 1960s in the United States and England (Laurian and Shaw 2009; Baker et al. 2010). In addition, the emergence of the strategy of enabling local communities in response to governments' failure to improve the conditions of informal settlements and poor neighbourhoods in developing countries (UN-Habitat

2003) can be considered as instances of practical developments in the last few decades. These developments generally address specific and limited concerns in certain areas rather than aiming to provide sustainable and comprehensive urban development. The other aspect of the emergence of such strategies is the inability of governments and formal institutions in promoting the objectives of urban development, without due regard to capacities and capabilities of local communities. Despite all these factors, the significant role of the state in urban development, whether at national or local level is undeniable. Without the ability, capacity, resources, and leadership of the state, the realisation of development objectives is very improbable, if not impossible (McGill 1998; Friedmann 1999; Evans et al. 2005, 2006; Hambelton and Gross 2007). In sum, the involvement and effectiveness of both local government and local community is necessary due to their complementary capacities as a prerequisite for sustainable development.

The other dimension of analysing the practical aspect of the relation between governance and sustainable development is the implementation of specific scientific and applied researches, for measuring the role of urban governance in the realisation of sustainable urban development. An important study in this field is the research for Developing Institutional and Social Capacities for Urban Sustainability (DISCUS) in the first half of the 2000s. The DISCUS research is based on the hypothesis that changing top-down approaches (government) to an approach based on discourse and dialogue (governance) enhances awareness, shared responsibility and acceptance among citizens, moving towards lower consumption of resources and more social inclusiveness. The research findings indicate that for making cities more sustainable, in addition to changing our approach from government to governance, adjustment and modification of the rules that cover the interactions between local governments and civil society are also necessary. Such changes are required for involving and strengthening people and the significant role of local governments in obtaining tangible, long-term results for sustainability (Evans et al. 2005, 2006).

In the DISCUS analysis model, the expected outcomes of sustainable development policies are introduced based on different modes of social capacity and institutional capacity (as the two main aspects of urban governance). According to this model, success in implementing sustainable development policies is dependent on the existence of high institutional and social capacities, while failure results from low institutional and social capacities. In addition, in case one of these capacities is high and the other is low, the result will be the partial progress of policies of sustainable development.

Analysing the relationship between urban governance and sustainable development, either in theoretical texts and policy documents or in practice indicates the decisive role of governance in promoting the objectives of sustainable development. In other words, mere emphasis on the actions and technical aspects of sustainability (focus on substantive dimensions) without due regard for procedural dimensions is not sufficient for realisation of the objectives of sustainable development. Following Evans et al. (2005) in the present research we have focused on an interpretation of governance that considers the role of local government in sustainable development through the evaluation of institutional capital and social capital for moving and progressing towards sustainable urban development in Iran.

12.3 Methodology

The chapter adopts a systematic review of meta-analysis approach and brings together results of a number of previous studies undertaken by the authors and other researchers on social and institutional development in cities of Iran more generally. In this regard, sustainability is considered as a dependent variable and procedural factors, including social and institutional capacities, are considered as independent variables.

The criteria for selecting research studies for the present study include the relevance and substantive relation, the general reliability of the research, and its time period. Regarding relevance, those research studies that have covered some factors of the theoretical model of the present research are included here. Regarding the general reliability of the research, considering the difficulty of determining the credibility of relevant resources for a systematic review, the research works are selected from approved Ph.D. dissertations, approved research in reliable research centres, papers published in reliable journals, and in a few cases, the reliable dissertations at Masters level. Finally, in accordance with research focus period the references were selected from researches conducted during the 1990s and later.

12.4 The State of Sustainability in Iranian Cities

As shown in Table 12.1 we have selected five research relevant projects on measuring sustainability conditions in Iranian cities. However, it should be noted that to reflect the concentration of political, economic, and cultural resources in the city, most reviews and researches are devoted to Tehran. Generally, these studies agree that environmental, social, and economic indices for the cities are undesirable or at most at intermediate level. While small in number, these findings support the researches conducted at national level by local and international institutes noted in the introduction (World Bank 2005; Ministry of Energy of Iran 2011; WGES 2009).

12.5 The State of Social Capital in Iranian Cities

The concept of social capital, like many other concepts and approaches including sustainable development and urban governance, as Sabatini (2008) puts it, evades definition. However, it can be considered from both subjective and objective

Table 12.1 Titles and characteristics of selected research for measuring sustainable development in Iranian Cities

Research references	Title	Research approach	Sample size	Significant findings
Mousa Kazemi Mohammadi (1999)	Evaluation of sustainable development in urban development: a case study on Ghom city	Quantitative	–	The statistical significance of differences and inequalities, the relation between urban issues and sustainability indices, the relation between population and unsustainability of development in different areas of the city
Tabibian and Faryadi (2001)	Evaluating the quality of urban environment in Tehran	Quantitative	–	The average quality of the total index of urban environment in Tehran in the year of the study (1996), including the ecological, social, economic, and cultural dimensions and indicators
Pourmoosavi (2005)	The security considerations of Tehran metropolis, based on indices of sustainable urban development	Quantitative	–	Security unsustainability in Tehran regarding ecological, economic, social, and cultural dimensions
Bararpour (2008)	Measuring the state of sustainability of local development in Kalardasht, using a strategic model	Quantitative	31	Unsustainable, poverty generating development: expansion of physical capital to the detriment of human and social capitals or erosion of natural capital
TURPC (2011)	Report on the state of environment in Tehran city	Quantitative	–	The gloomy condition of the environment in Tehran, in the areas that determine the environment quality, including air, water, soil, bio-diversity, natural disasters, waste disposal and human habitat

perspectives. This division is the focal point of the research that are reviewed here as well, and can be found in related texts (such as Putnam 2000; Mandarano et al. 2010; Andrews 2011) with expressions such as attitudinal and structural, or normative and network. The subjective dimension of social capital relates to norms, attitudes and levels of social trust, and the objective dimension is crystallised in networks, relations, formal and informal participation. Table 12.2, shows the 10 selected studies.

The results of selected research works in national and urban surveys suggest that as a whole, despite variations, trust at macro and medium levels is relatively low. The other dimension of social capital related with the objective dimension and the quality of relations and social networks, measured in terms of participation, indicates that the condition of this dimension is even worse than the subjective dimension. For instance, in Firouzabadi (2005) and Aghili et al. (2009) who evaluate the state of trust at macro-level as almost medium, the extent of formal participation and membership in civil institutions have been very low.

12.6 The State of Institutional Capital in Governing Iranian Cities

The institutional capacity, in the sense used by the authors and researchers of collaborative planning discourse, refers to the abilities and capacities of state and executive institutions and organisations for responsiveness and management of social and environmental challenges through decision-making and planning processes. In this discourse, institutional capacity includes cases such as organisational division and fractures in the distribution of power, and within a specific framework, includes knowledge resources, relational resources, and mobilisation capacity of resources (Polk 2011). From this specific perspective, no research was conducted for analysing the relation between institutional capacity and sustainable urban development in Iran. However, the development of institutional capacity, if considered as the concept of enabling local and urban management for confronting different urban and social challenges, has been one of the controversial issues in recent years in the field of urban management in Iran. Table 12.3 shows titles and characteristics of selected research on Institutional Capital in Governing Iranian Cities.

A review of these researches in shows the existence of numerous fundamental problems in decision-making and urban management systems. In addition, the weakness of participation and governance in the urban management system (Barakpou 2002; Iranian Management Services Co. 2009) are also identified as key obstacles to more effective institutional capacity.

Table 12.2 Titles and characteristics of selected research on social capital in Iran

Research references	Title	Approach	Sample size	Major findings
The Iranian Ministry of Culture and Islamic Guidance (2001)	National survey of Iranians' values and attitudes	Quantitative	16,800	High to medium levels of trust in personal and family relations. Low to medium level trust in professional and state institutions.
The Iranian Ministry of Culture and Islamic Guidance (2003)	National survey of Iranians' cultural behaviour	Quantitative	4540	same as above
Yazdanpanah (2003)	Reviewing the factors affecting the extent of participation of citizens over 18 in 22 areas of Tehran	Quantitative	880	Low level of social participation of the respondents (54.4 % 'never' and 41.9 % 'low'); A sense of personal powerlessness and vagueness in 78.2 % of respondents
Firouzabadi (2005)	Social capital and effective factors on its formation in Tehran city	Quantitative	1759	Four main elements of social capital: general trust, institutional trust, awareness and formal participations, informal participations; average social capital: 4.78 out of 7
Zeinabadi (2008)	Reviewing the state of trust in the Iranian society and ways to restore it	Quantitative	–	Relatively desirable state of trust at micro-level (trust to family members, relatives and friends), and undesirable condition of trust in medium level (trust to different guilds), and macro-level (trust to government and public institutions)

(continued)

Table 12.2 (continued)

Research references	Title	Approach	Sample size	Major findings
Hoseini et al. (2007)	Ranking social capital in the capitals of provinces in Iran	Quantitative	–	Tehran city has the worst and Ilam the best average of all six types of social capital
Calabrese et al. (2008)	Environmental policies and strategic communication in Iran	Quantitative	1200	The environmental concerns are not a priority for the majority of respondents; respondents believe that the government is responsible for reducing the environmental problems at no cost for the people
Aghili et al. (2009)	Social capital and responsible environmental behaviours in North of Iran: a case study in Gilan, Mazandaran, and Golestan provinces	Quantitative	750	The average index of social capital is slightly above average (2.838 out of 5) and the index of responsible environmental behaviours is above average (3.5 out of 5)
Zolfaghari and Fe'li (2009)	Reviewing the effects of social status of individuals on their attitude towards democratic values; a case study of the young inhabitants of Tehran, with ages ranging from 18 to 29	Quantitative	391	Identifies a reverse relation between social capital and tendency to democratic values; low social capital in the sample
Zare et al. (2010)	The role of social capital in citizens' participation in works related with forest parks: a case study of Tehran city	Quantitative	205	Just over half of respondents have medium level of social capital (20 % low, 54.6 % medium, and 25.4 % high)

Table 12.3 Titles and characteristics of selected research on institutional capital in governing Iranian Cities

Research references	Title	Approach	Sample size	Major findings
Kazemian (1994)	Designing appropriate urban management system for Iranian cities; a case study: Mashhad	Qualitative	–	Unclear status and identity of urban management in the macro-management of the country, lack of elected elements and institutions of citizens, serious and full scale dependence on central government and conflicts between organisations and elements
Barakpou (2002)	Transition from urban government to urban governance in Iran: a case study of Hamedan and Eslamshahr	Quantitative	307	Low final score regarding criteria of urban governance in Eslamshahr (364 out of 1000) and Hamedan (342 out of 1000)
Kazemian and Saeedi Rezvani (2001–2003)	Feasibility study for devolution of new responsibilities to municipalities	Qualitative	–	Limitation of municipalities' responsibilities in Iran as compared with other countries; impossibility of official, legal interference of municipalities in policy-making and planning; necessity of adopting an approach aimed at refining the responsibilities of municipalities
Urban Processing and Planning Co. (2003)	The strategic-comparative studies of world cities and Tehran	–	–	Lack of integrated urban management; limiting Tehran municipality to some urban services; inattention to areas of economic, social, tourism, and environmental development

(continued)

Table 12.3 (continued)

Research references	Title	Approach	Sample size	Major findings
Iranian Management Services Co. (2005)	Restructuring the urban management and finance system in Tehran metropolis	Qualitative	–	Interference of political fields and urban management limits; unclear relations between the central government and municipalities; lack of strong private organisations in production and delivery of urban services; weak public participation; dependence of municipalities on unsustainable revenues
Akhoundi et al. (2008)	Pathology of urban governing model in Iran	Qualitative	–	Lack of separation of political leadership from technical management; limitation of the duties of urban management institutions; adoption of a similar urban management model in all cities; weak check and balance mechanism for the relations between the council and the municipality
Pourzarandi et al. (2007)	Comprehensive plan for sustainable revenues for Tehran municipality	Quantitative	–	Heavy reliance of Tehran municipality on unsustainable revenues
Iranian Management Services Co. (2009)	Institutional review and designing training programs in Tabriz, Bandar Abbass, Kermanshah, Zahedan, and Sanandaj Municipalities	Qualitative	–	Shortage of capacity in selected municipalities in areas of planning and coordination, institutional arrangements, financial resources and revenues, attracting public participation

12.7 Conclusion

We started this chapter by applying Faludi's (1973) concepts on substantive and procedural issues in planning sustainable urban development. We have utilised these concepts to undertake a systematic review of existing research on these issues in the context of Iran. In the main much of the examined research has not directly addressed the core concerns of this review in terms of applying social and institutional capital to urban sustainability. Nevertheless, the existing evidence allows us to make the following concluding observations.

- Regarding the substantive dimension or the state of sustainability of urban development, the reviewed research and reports indicate the unsustainable state of development at national and local levels in Iran. Such findings are compatible with findings of other research showing the unsustainability of development in developing countries, especially their large cities (UN-Habitat 2009; Ojendal and Dellnas 2010).
- Regarding procedural dimension and the effective factors in development sustainability, we can note the following:
  - The findings indicate low levels of social and institutional capacities in the society. Therefore and merely based on previous research, a direct correlation between low social and institutional capital and low sustainability of development can be observed in Iranian cities. In the meantime, due to shortage of previous research, we have had to consider social capital and institutional capital at a more general level in this review. However, their relations with indices of sustainable development were not studied.
  - Regarding social capital, the majority of research studies at national and urban levels, indicate desirable degrees of trust at micro-level (relations with the family, relatives, and friends) and undesirable degrees at medium and macro-level, i.e., trust in government and formal organisations. Meanwhile, that part of trust that paves the way for social participation is the trust at levels beyond personal level and especially trust in formal organisations and institutions.
  - Regarding the institutional capital, the findings of the majority of the reviewed researches suggest functional fragmentation and lack of integrity in urban and local government in Iran. The lack of integrity and authority of urban governance results in a waste of resources, reduction of effectiveness of activities and a lack of accountability of different institutions. In addition, we can mention the weakness of the roles of public and elected institutions in fundamental areas such as economic development and employment, social welfare including education, security and public health, and urban planning and development. These deficiencies become particularly important in terms

of execution of power and policy making and reduce the role of public institutions to discharging routine duties and services.
  - In addition, the low level of public participation indices, whether in research related with social capital or in research on institutional capital, is another challenge to sustainable urban development in Iran.

- The above findings confirm the theoretical model of this research that argues where the level of social capital and institutional capital are low, policies of sustainable development are likely to fail. In Iran, it was mentioned that in the last two decades, almost all important policy documents such as the Vision of Development, five-year plans for economic, social, and cultural development, national regulations for land use planning, development visions and charters for different cities and regions of the country, have frequently emphasised the necessity of sustainable development. However, different evaluations at national and urban level, a significant part of which reviewed here, indicate the failure of sustainable development policies.
- In the planning environment of Iran, the main actors and elements of urban governance are the state (as a sectorial, appointed institution), urban management (an inter-sectorial, elected institution), and the urban civil society. The type of relation and interaction among these elements, determine the quality of governance, and in accordance with theoretical foundations of this research, the quality of urban development. Based on the operational definitions used in the research studies reviewed here, social capital is mainly crystallised in themes such as trust (at different micro, medium, and macro levels) and citizen participation. The institutional capital hinges upon creating coordination and cooperation between public sector (urban management) and state sector as summarised in the concept of integration. Therefore, concepts of trust, participation, and integration in interactions among the state, the elected public institution, and the urban society, comprise the key themes in all research in the field of the relation between urban governance and sustainable development.
- In the present research, the relation between urban governance and sustainable development in Iran was studied based on previous research in which elements of governance (institutional capital and social capital) and sustainability are addressed in separation. Nevertheless, this chapter provides a preliminary framework that clarifies the conceptual relation between these elements and their different dimensions (Fig. 12.1). Some noteworthy themes in the field of the relation between urban governance and urban sustainable development are as follows:

  - The contribution and role of the state and urban management in promoting policies of sustainable development.
  - The extent of efficiency of state and urban management structures in promoting policies of sustainable development.
  - The factors and backgrounds that may be effective in expanding social capacity (trust and participation) for sustainable development.
  - Grounds for expansion of institutional capacity for sustainable development.

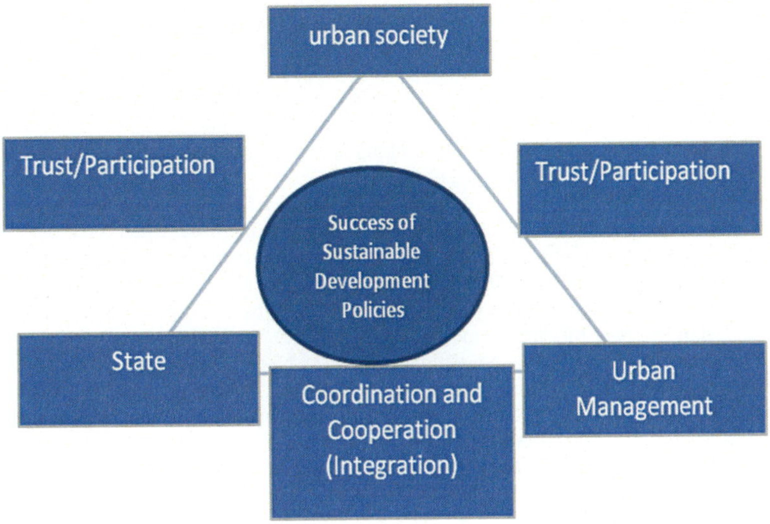

Fig. 12.1 The preliminary framework of the conceptual relation between urban governance and urban sustainable development and their different dimensions (authors)

References

Aghili M, Khoshfar G, Salehi S (2009) Social capital and responsible environmental behaviors in North of Iran: a case study in Gilan, Mazandaran, and Golestan provinces. Oloome Keshavarzi va Manabe Tabiei (in Persian), 16(1)

Akhoundi AA, Barakpou N, Asadie I, Taherkhani H, Basirat M (2008) Pathology of urban governing model in Iran. Pajooheshhaye Joghrafiaee (in Persian) 63

Andrews R (2011) Social capital and public service performance: a review of the evidence. Public Policy Adm 27(1):49–67

Baker M, Hincks S, Sherriff G (2010) Getting involved in plan making: participation and stakeholder involvement in local and regional spatial strategies in England. Environ Plann C: Gov Policy 28:574–594

Barakpou N (2002) Transition from urban government to urban governance in Iran: a case study of Hamedan and Eslamshahr. Unpublished Ph.D. dissertation, University of Tehran, Tehran

Bararpour K (2008) Measuring the state of sustainability of local development in Kalardasht, using a strategic model. Pajooheshhaye Joghrafiaee (in Persian) 63:173–192

Calabrese D, Kalantari K, Santucci FM, Stanghellini E (2008) Environmental policies and strategic communication in Iran. The World Bank, Washington, DC

Evans B, Joas M, Sundback S, Theobald K (2005) Governing sustainable cities. Earthscan, London

Evans B, Joas M, Sundback S, Theobald K (2006) Governing local sustainability. J Environ Plann Manage 49(6):849–867

Faludi A (1973) A reader in planning theory. Pergamon Press, Oxford

Firouzabadi A (2005) Socia capital and effective factors on its formation in Tehran city. Unpublished Ph.D. dissertation, University of Tehran, Tehran

Friedmann J (1999) The common good: assessing the performance of cities. In: Friedmann J (ed) Urban and regional governance in the Asia Pacific. Institute of Asian Research, Vancouver, pp 1–16

Hambelton R, Gross JS (2007) From governance to governing. In: Hambleton R, Gross JS (eds) Governing cities in a global era. Palgrave Macmillan, New York, pp 213–224

Hoseini AH, Elmi Z, Sharepour M (2007) Ranking social capital in the capitals of provinces in Iran. Refahe Ejtemaee (in Persian) 7(26):59–84

IEA (International Energy Agency) (2010) Key world energy statistics. [online]. Available from http://www.iea.org/textbase/nppdf/free/2010/key_stats_2010.pdf. Accessed May 2012

IEA (International Energy Statistics) (2011) Key world energy statistics. [online]. Available from http://www.iea.org/textbase/nppdf/free/2011/key_world_energy_stats.pdf. Accessed May 2012

Iran Ministry of Culture and Islamic Guidance (2001) National survey of Iranians' values and attitudes. Iran Ministry of Culture and Islamic Guidance, Tehran

Iran Ministry of Culture and Islamic Guidance (2003) National survey of Iranians' cultural behavior. Iran Ministry of Culture and Islamic Guidance, Tehran

Iranian Management Services Co (2005) Restructuring the urban management and finance system in Tehran metropolis. Iranian Management Services Co, Tehran

Iranian Management Services Co (2009) Institutional review and designing training programs in Tabriz, Bandar Abbass, Kermanshah, Zahedan, and Sanandaj Municipalities. Iranian Management Services Co, Tehran

Kazemian G (1994) Designing appropriate urban management system for Iranian cities; a case study: Mashhad. Unpublished Master's thesis, University of Shahid Beheshti, Tehran

Kazemian G, Saeedi Rezvani N (2001–2003) Feasibility study for devolution of new responsibilities to municipalities. State Organisation of Municipalities and Districts, Tehran

Keivani R (2009) A review of the main challenges to urban sustainability. Int J Urban Sustain Dev 1(1–2):5–16

Laurian L, Shaw MM (2009) Evaluation of public participation: the practices of certified planners. J Plann Edu Res 28:293–309

Mandarano L, Meenar M, Steins C (2010) Building social capital in the digital age of civic engagement. J Plann Lit 25(2):123–135

McGill R (1998) Urban management in developing countries. Cities 15(6):463–471

Ministry of Energy of Iran (2011) The balance sheet of energy. Ministry of Energy of Iran, Tehran

Mousa Kazemi Mohammadi M (1999) Evaluation of sustainable development in urban development: a case study on Ghom city. Unpublished Ph.D. sissertation, University of Tarbiat Modarres, Tehran

MRUDI (Ministry of Roads and Urban Development of Iran). (n.d.). [online] Available from http://www.udro.org.ir/. Accessed May 2012

Ojendal J, Dellnas A (2010) Governance dilemmas of sustainable cities. ICLD

Polk M (2011) Sustainability in practice: the interpretation of sustainable development in a regional planning arena for dialogue and learning in Western Sweden. Plann Theor Pract 11 (4):481–497

Pourmoosavi M (2005) Study of security unsustainabilities of Tehran metropolis, based on indices of sustainable urban development. Unpublished Ph.D. dissertation, University of Tehran, Tehran

Pourzarandi HM, Hashemi SM, Taherkhani H (2007) Comprehensive plan for sustainable revenues for Tehran municipality. Tehran Municipality, Tehran

Urban Processing and Planning Co. (2003) The strategic-comparative studies of world cities and Tehran. Tehran Municipality, Tehran

Putnam RD (2000) Bowling alone: the collapse and revival of American community. Simon & Schuster, New York

Rogers PP, Jalal KF, Boyd JA (2008) An introduction to sustainable development. Earthscan, London

Sabatini F (2008) Social capital as social networks: a new framework for measurement and an empirical analysis of its determinants and consequences. J Socio-economics 38:429–442

Statistical Centre of Iran (2012) [online]. Available from http://www.amar.org.ir/. Accessed May 2012

Tabibian M, Faryadi S (2001) Evaluating the quality of urban environment in Tehran. MohitShenasi (in Persian) 28:1–12

TURPC (2011) Tehran state of environment. Municipality of Tehran, Tehran

UN-Habitat (2003) The challenges of slums. Earthscan, London

UN-Habitat (2009) Planning sustainable cities. Earthscan, London

WGES (2009) Report on environmental sustainability plan. Tehran

World Bank (2005) Islamic republic of Iran: cost assessment of environmental degradation. Report No. 32043-IR, World Bank

Yazdanpanah L (2003) Reviewing the factors affecting the extent of participation of citizens over 18 in 22 areas of Tehran. Unpublished Ph.D. dissertation, University of Allame Tabatabaee, Tehran

Zar'e S, Namiranian M, Fami HS, Ghasemi J (2010) The role of social capital in citizens' participation in works related with forest parks: a case study of Tehran city. Jangal Iran (in Persian) 2(4):273–285

Zeinabadi M (2008) Reviewing the state of trust in Iranian society and ways to restore it. Pajooheshname (in Persian) (16):33–65

Zolfaghari A, Fe'li J (2009) Reviewing the effects of social status of individuals on their attitude towards democratic values; a case study of the young inhabitants of Tehran, with ages ranging from 18 to 29. Pajoohesh Ejtemaee (in Persian) 2(5):171–201

Chapter 13
Tehran: Old and Emerging Spatial Divides

Hamidreza Rabiei-Dastjerdi and Maryam Kazemi

Abstract Social and spatial inequalities are important challenges in today's urban life. Socio-economic inequalities sometimes manifest themselves in the form of spatial inequality. In other words, spatial inequality represents social inequality or distribution disorder in the social system. Tehran, as a metropolis and the capital of Iran, has a crucial role in the national system. The continuous growth and development of Tehran has unbalanced its system. The city has attracted a large population. As a centre, Tehran plays an important role in the country's macro socio-economy but its urban components suffer from spatial inequality and unbalanced distribution of services. One of the most visible characteristics of Tehran is its North-South spatial inequality and polarisation created by modernisation and its elements such as cars and newly emerged business districts. Although the revolution of 1979 came with a promise of equality and improvement of lifestyle for the poor and the marginalised, the growing gap between the rich and the poor and spatial inequality in Tehran in the ensuing years have been widened. Although most urban planners and sociologists in Iran think that spatial inequality in Tehran is a historic trend and not caused by globalisation, others believe that globalisation has intensified it. A global city or not, this chapter illustrates old and new emerging spatial inequality lines and patterns in Tehran, and also examine the theory of 'Global City' (Saskia Sassen's theory) between 1996 and 2006 on this case.

Keywords Global city · Polarisation · Spatial inequality · Tehran

H. Rabiei-Dastjerdi (✉)
Politecnico di Milano, Milano, Italy
e-mail: hamidreza.rabiei@polimi.it

M. Kazemi
Accademia delle Belle Arti di Brera, Milano, Italy
e-mail: maryamkazemi49@gmail.com

F.F. Arefian and S.H.I. Moeini (eds.), *Urban Change in Iran*,
The Urban Book Series, DOI 10.1007/978-3-319-26115-7_13

Abbreviations

GaWC Globalization and World Cities Research Network
MENA Middle East and North Africa CBD
GDP Gross Domestic Product
MAUP Modifiable Areal Unit Problem

13.1 Introduction

The global city is one of the controversial concepts of globalisation including: international flows of information, goods, labour and capital with different functions and domains, and also political, cultural, and Infrastructural characteristics in the global economic system. Global cities suffer from growing social and spatial inequality, as Sassen's (1991) studies of three most important and different global cities of the world in three continents (London, New York and Tokyo) show. Castells (1989: 343) and Sassen (1991) believe that the observed social polarisation (Sassen) or new urban dualism (Castells) pattern in New York and Los Angeles could be generalised to other global cities of the world.

Social polarisation is an important challenge in today's urban life in global cities. Socio-economic polarisation and inequalities occur in the form of spatial inequality (Castells 1989: 227). In other words, spatial inequality is considered as a kind of social inequality or distribution disorder in the social system that highlights the unequal distribution of opportunities and positions in space. With regard to damages caused by spatial inequality, one can point to a loss of optimal use of social space, along with other social inequalities, the geographical concentration of poverty and intensification of segregation and polarisation, the weakening of spatial and social coherence, the prevention of optimum allocation of urban facilities and services as well as an imbalanced population distribution through intensifying migrations.

As mentioned earlier, spatial inequality in Tehran has historical backgrounds, while Sassen's social polarisation theory (1991) in global cities indicates that social structure changes are consequences of economic structure changes, it means service sector grows while production sector shrinks; these cities attract both high and low paid jobs that increases social and economic distance between people. This chapter tries to find the relationship between spatial inequality and social disparity and injustice. It also illustrates spatial inequality and translates social polarisation to spatial inequality using maps in Tehran.

13.2 Tehran: A Global City or Not?

Tehran has been the capital of Iran since 1795, with a population of 8,429,807 (Iran Statistical Centre 2012). It is also Iran's largest city, the centre of the country's largest urban area, one of the largest cities in Western Asia, and one of the largest cities in the world. Tehran is also the economic centre of Iran. Many of Iran's public-sector workforce and a large number of industrial firms are located in Tehran, and almost half of these workers work for the Government.

Considering Tehran as a global city is a controversial and challenging issue. Peter Taylor (GaWC) listed Tehran as a global city in 1998 and deleted it from the list in 2004. Also, Short (2004) called Tehran a 'black hole' in the global urban network. 'Global Cities are big, but not all big cities are global cities' (Ibid). In this field, it is arguable that Tehran has invisible and semi-visible connections to many other global cities such as Dubai, Istanbul, and Los Angeles in global network cities. These visible and semi-visible connections are remarkable because of many factors, namely: political, historic, economic, and cultural context and background. In other words, Short did not see the other side of the coin in this black hole, namely Tehran's connection with the global cities network, in spite of his interesting quantitative analytical approach. It could be the subject of a different research.

13.3 Spatial Inequality in Tehran

This section reviews some available indices to present and visualise spatial inequality at urban level and tries to analyse these indices, although, there are other effective macro factors which are important for the analysis of spatial inequality and disparities in Tehran. At the macro level, the effects of geography, economy, and history, and also effects of economic and political strategies should be taken into consideration on urban socio-spatial inequalities (World Bank 2010).

The result of macro factors in the Middle East and North Africa (MENA) is that social and spatial inequality is inevitable because all areas do not have equal potentials, resources, and opportunities such as water and security for a balanced development. The most important dynamic consequence of inequality is uneven development and a high spatial concentration of economic and human capital in Tehran that causes many social problems such as migration to Tehran and the concentration of poverty, as this concentration of facilities and capital attract people from different parts of the country, not only white-collar but also blue-collar workers. In fact, social disparity is a simple and understandable phenomenon. 'It is the gaps in living standards between people in different places' (World Bank 2010). People come to big cities such as Tehran to fill this gap.

Among many key data sets, seven parameters (land price, family size, population density, recreational spaces, housing ownership, the number of managers and experts, and the number of unskilled workers) are chosen here as most relevant indices to show social and spatial inequality and injustice in different parts of the city from the most up-to-date available data from the Tehran Municipality, Atlas of Tehran Metropolis (2006). It is worth mentioning that Tehran is divided into 22 districts with every district having its own administrative centre (Fig. 13.1).

Figure 13.2 is a map which presents land prices in Tehran. The map divides prices into six classes, with darker reds indicating higher land prices in the city. The difference in land property value reflects the income bracket of residents. This shows six categories in which north-south stratification is clear. The property value decreases southward with the vast difference between the highest price in districts 1 and 3 and the lowest in 18, 19, and 20. It is acceptable that the land price of an area has a direct relation with other economic factors as well as with facilities and services.

Figure 13.3 portrays family size in Tehran, and divides it into four classes, with darker colours showing larger family sizes, and showing southern parts as housing larger families and central parts smaller ones. Also, there are three hotspots in the north (marked with A, B, and C), where new large-scale residential projects, mostly for governmental employees, are built. Southern parts are more affordable for

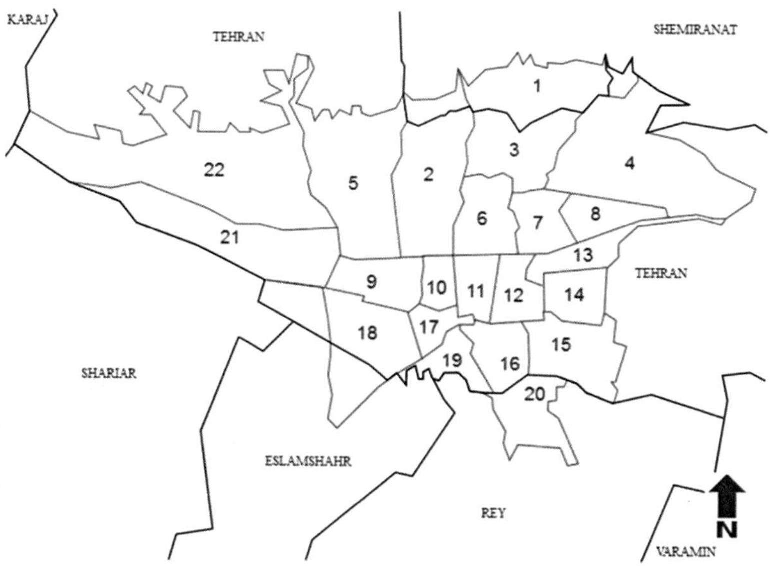

Fig. 13.1 Administrative map of Tehran (Atlas of Tehran Metropolis 2006). Reproduced by permission of Tehran Municipality ICT Organisation

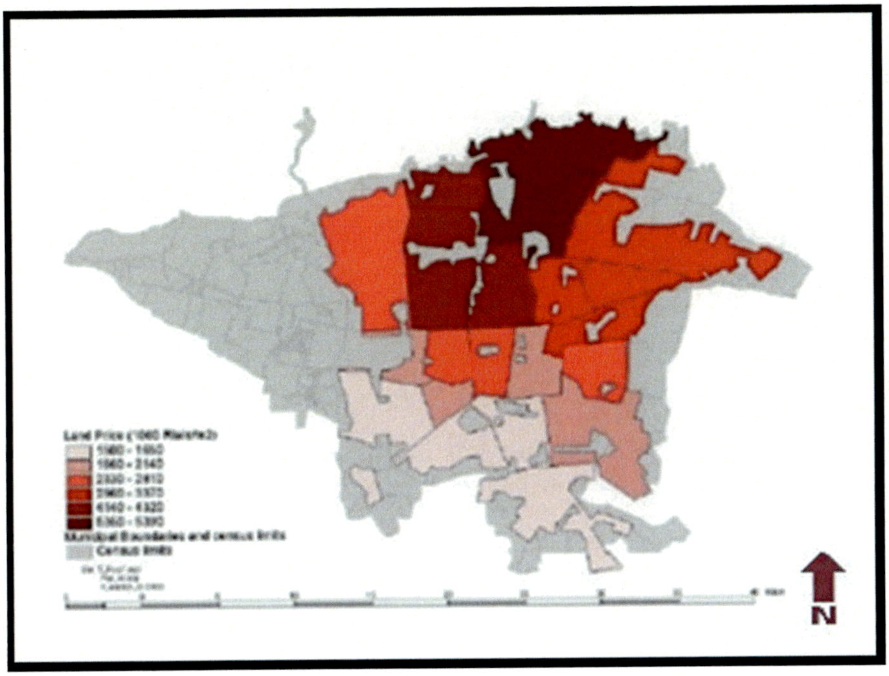

Fig. 13.2 Land price (Atlas of Tehran Metropolis 2006). Reproduced by permission of Tehran Municipality ICT Organisation

migrants from different parts of the country because of their lower land and rent prices. These newcomers usually have bigger family sizes. In addition, there is a higher birth rate than other parts here because of the lower birth control and education level.

Figure 13.4 presents population density, classified in five classes, follows the same pattern as family size. Districts 4 and 5 are parts of the city that attracted more people during the last two decades and to some extent populated. Land prices have dramatically increased during recent years in these districts.

Figure 13.5 which is classified in three classes, demonstrates that northern parts (1, 2, 3, 5 districts) as enjoying more recreational spaces than other districts. It has a meaningful relation with land price map representing a social injustice in different areas of Tehran as well.

Figure 13.6 presents housing ownership in four classes; it shows that in the northern part most people are house owners while in the southern part houses are predominantly rented. The anomaly in eastern and western parts are because of development projects in which land prices were not high not long ago before new housing projects.

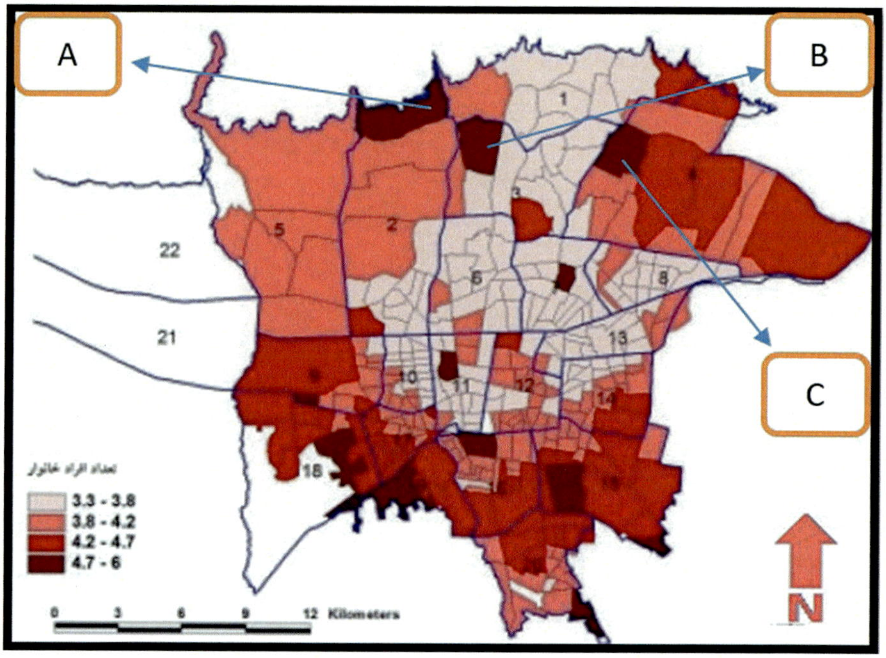

Fig. 13.3 Family size (Atlas of Tehran Metropolis 2006). Reproduced by permission of Tehran Municipality ICT Organisation

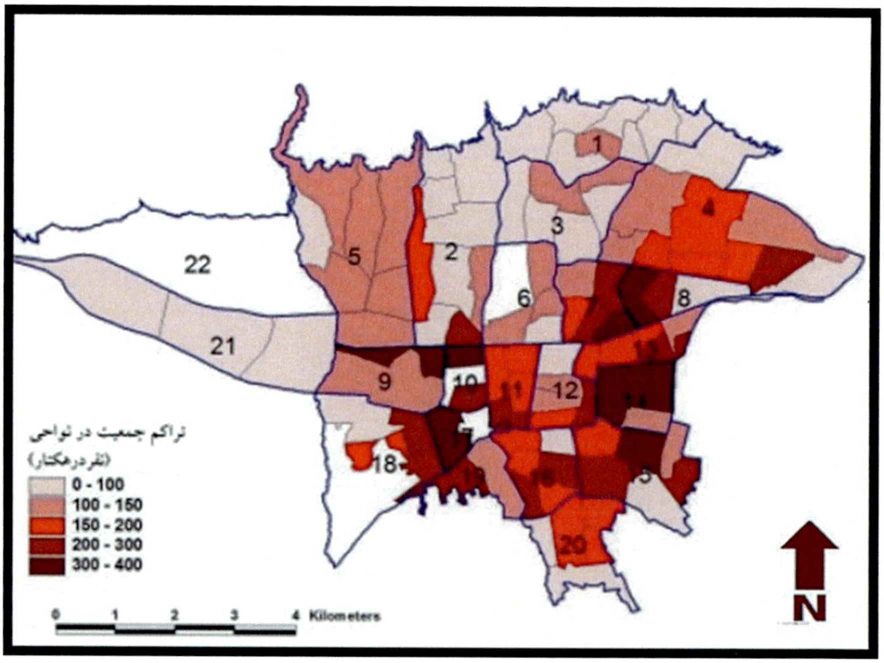

Fig. 13.4 Population density (Atlas of Tehran Metropolis 2006). Reproduced by permission of Tehran Municipality ICT Organisation

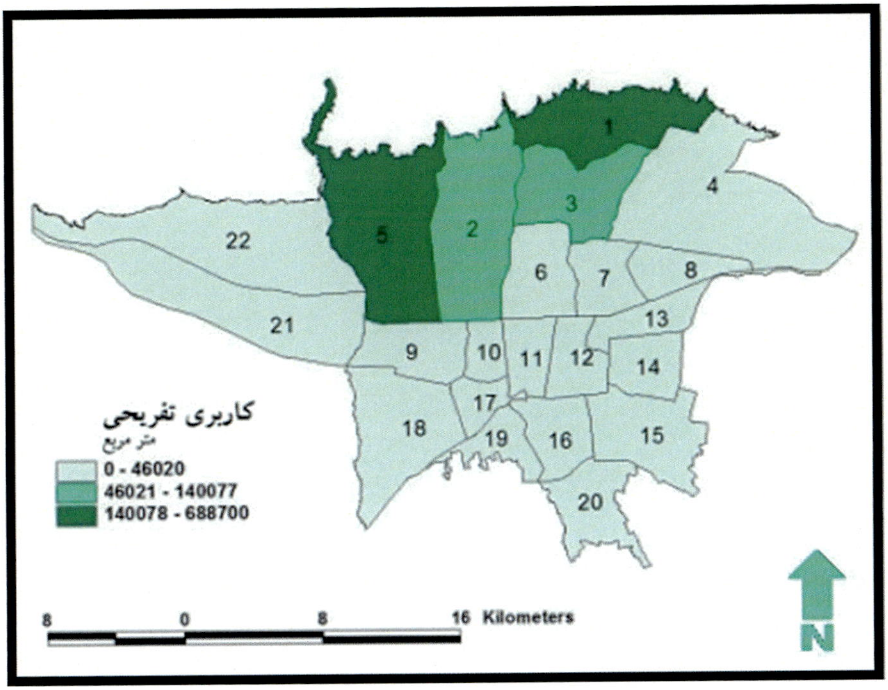

Fig. 13.5 Recreational spaces (Atlas of Tehran Metropolis 2006). Reproduced by permission of Tehran Municipality ICT Organisation

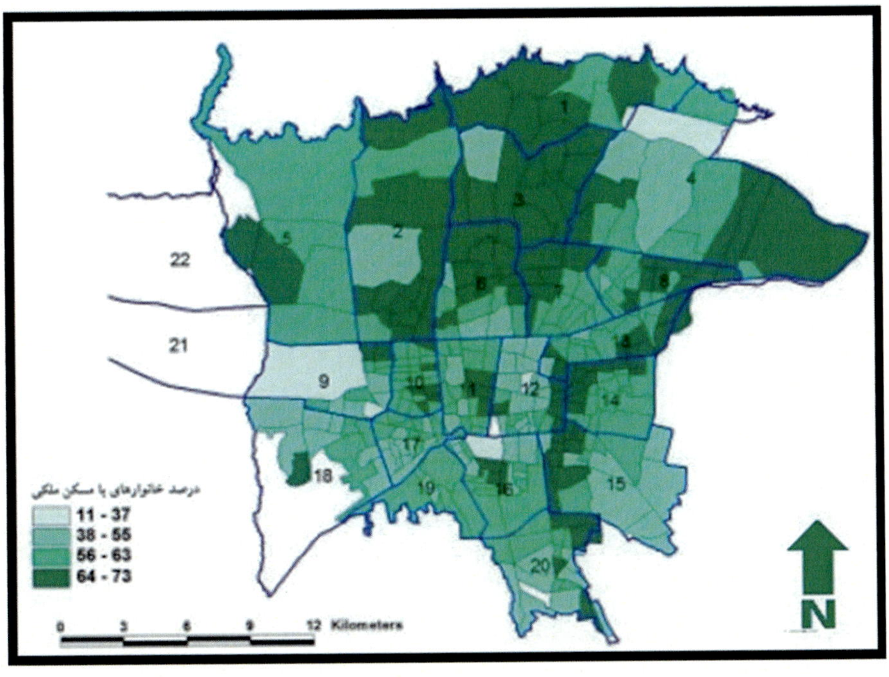

Fig. 13.6 Housing ownership (Atlas of Tehran Metropolis 2006). Reproduced by permission of Tehran Municipality ICT Organisation

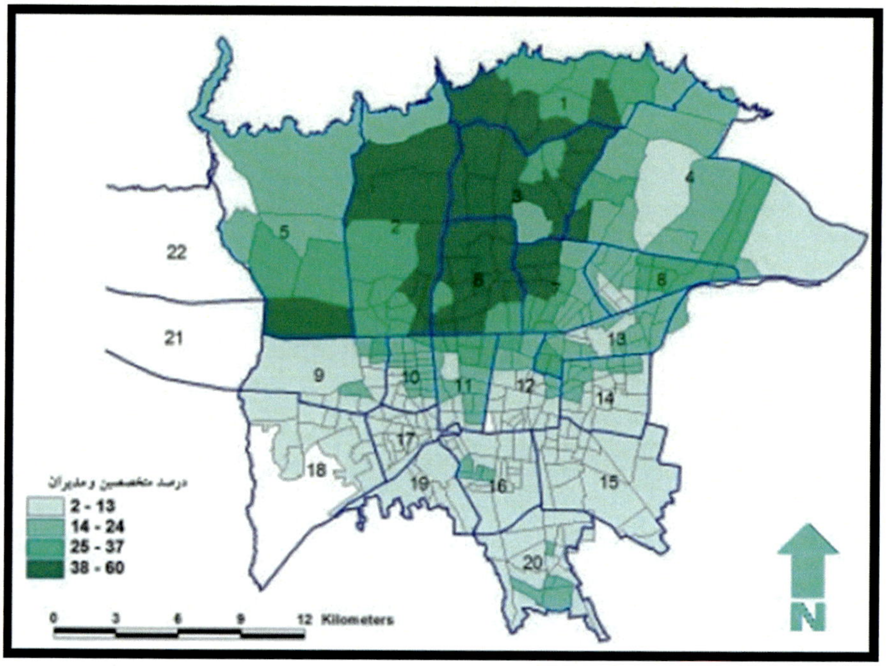

Fig. 13.7 Numbers of managers and experts (percent) (Atlas of Tehran Metropolis 2006). Reproduced by permission of Tehran Municipality ICT Organisation

Figure 13.7 is the statistics of highly-skilled workers classified into four classes based on the percentages of managers and experts living in those areas. It reflects the education level as well, and illustrates a visible spatial gap between well-educated people and unskilled workers in Tehran. Moreover, the data here is consistent with those on land price, housing ownership, and family size maps.

Figure 13.8 indicates the percentages of unskilled workers in four classes. This index is another indicator of spatial and social inequality in Tehran despite the abnormal density on the northern part, which can be explained by jobs that do not need specific skills and education, for example restaurant jobs, car-washing, and gardening. These workers try to keep their distance with the places where they work; but to save the cost of commuting they often live at their workplaces for free. The contrast between this map and that of land prices is interesting, but most of these dwellers and workers are living in hidden small slums in northern parts. They live in their workplaces such as restaurants, gardens, and workshops with minimal or no sanitary facilities and substandard amenities; this hypothesis, however, would take further studies to prove. It is comparable with Singapore and Dubai hidden slums. Also, there are a few old villages in the northern part of the city that are part

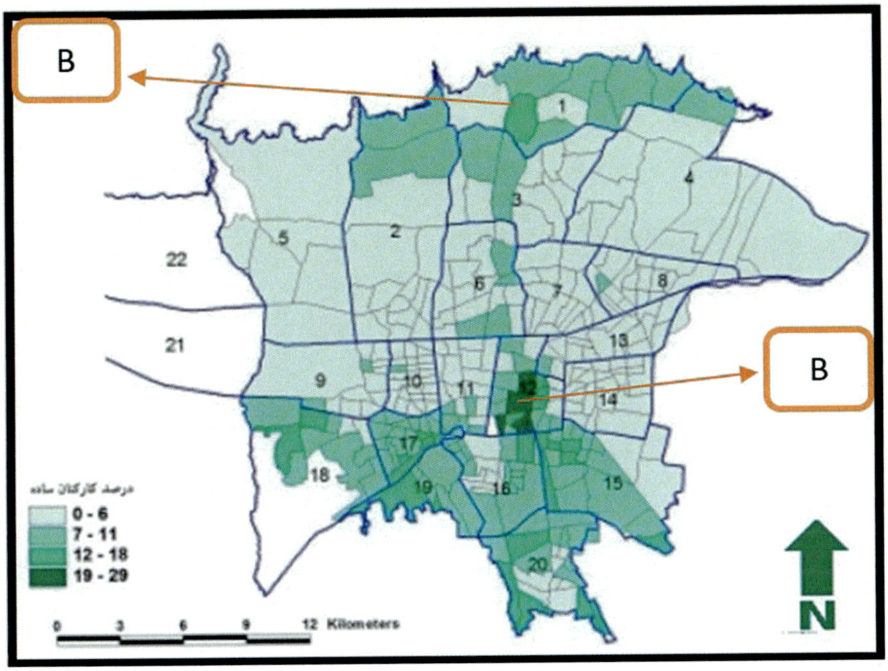

Fig. 13.8 Number of unskilled workers (percent) (Atlas of Tehran Metropolis 2006). Reproduced by permission of Tehran Municipality ICT Organisation

of it now, but with different fabric, people, and housing conditions (sometimes, even their accents are different from Tehranis) such as Zargandeh, Deh-e-Vanak, and Rostam Abad. These people are similar to what Ganes (1982) calls urban villagers. In other words, they are villages within a city (urban villages) in which unskilled workers live. These villages have been annexed to Tehran during the last century. In addition, There is a hotspot in the northern part: the Tajirsh old Bazaar and its surroundings; and also in the southern part, at a yet higher density than its deprived neighbourhood, at the grand Bazaar of Tehran.

Figure 13.9 illustrates the population growth rates in five classes. At first glance, this map is in contrast with north-south spatial inequality divide as it shows almost the same population growth rate in northern and southern parts with the lowest rate in central Tehran.

This can be explained by three phenomena; first: hidden slums (1, 2, 3, and 6 districts). District 6 is the central business district of Tehran (CBD); in this district, home-based office is a common new phenomenon even for well-educated people. The second is newly developed areas (5, 22 and 4), and third the emerging new west-east spatial inequality in Tehran.

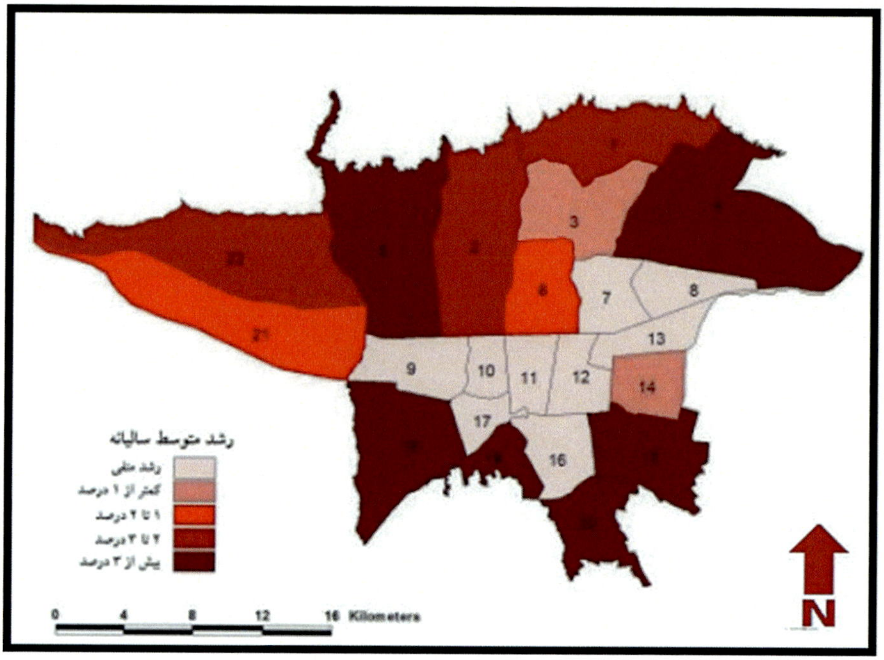

Fig. 13.9 Population growth rate (percent) (Atlas of Tehran Metropolis 2006). Reproduced by permission of Tehran Municipality ICT Organisation

13.4 Spatial Inequality Analysis and Discussion

The spatial inequality analysis method here is based on Sassen's theory of social polarisation in global cities, in order to raise the awareness of the city management about problems of social and spatial inequalities. It is worth mentioning that the scale of maps is 1:250,000, and the data has been integrated according to this scale.

13.4.1 Occupation Structure Change

Unfortunately, there is no reliable income data in Tehran, but the following table consists of occupation structure change, driven from two surveys in Tehran 1996 and 2006.

Table 13.1 indicates that the number of managers and experts (upper strata) and unskilled workers (lower strata) increased in Tehran this decade. But the services sector did not change while, according to the global city theory, it must have increased. On the other hand, according to Iran Statistics Centre data, the share of production has decreased in all cities and the whole country, while services share

Table 13.1 Occupation structure change in Tehran, 1996 and 2006 (Iran Statistical Centre, Tehran occupations statistics, 1996 and 2006 surveys, dbf format, originally 42 and 36 classes summarised into five classes by authors)

Occupation	Total number year: 1996	Total number year: 2006	Percentage year: 1996	Percentage year: 2006	Change
Manager and experts	681,800	1,173,377	18.94	22.32	+3.38
Services	1,664,126	2,448,722	46.23	46.58	+0.35
Production	810,688	959,348	22.52	18.25	−4.27
Unskilled workers	63,384	315,172	1.76	6.00	+4.24
Unknown	379,935	361,602	10.55	6.85	−3.7
Total	3,599,933	5,258,221	100.00	100.00	0.00

has increased. Moreover, gross domestic product (GDP) has decreased dramatically during 2000 to 2009, (according to World Bank reports), exacerbated by Western sanctions against Iran and the global economic crisis.

13.4.2 Lifestyle

The Iranian lifestyle is less individualistic compared to the West; with some trying to keep their relation and spatial and geographical proximity with their family, friends and their spatial memories of their birth and growing place as well what Yi-Fu Tuan called Topophilia (Tuan 1974). Topophilia is strong emotional attachment of people to the place they live in, with different aspects such as cultural identity, and personal and collective memories. Sometimes an affluent person lives in a deprived and poor locality just to be near his or her family and friends. It means that the geographical position of a person or family could not be the only evidence explaining the socioeconomic conditions of all people who live in that area. This issue could add to the problem of interpreting the maps.

13.4.3 Modifiable Areal Unit Problem (MAUP)

We used categorical maps to illustrate spatial inequality in Tehran. Modifiable Areal Unit Problem (MAUP) is an important and critical issue here: namely, that 'the interpretation of a geographical phenomenon within a map depends on the scale and partitioning of the areal units that are imposed on the map' (Hayward and Parent 2009). MAUP has two main aspects: the scale and the zoning. The problem of the scale is that using different scales possibly could reveal different patterns and facts pertained to existing social and spatial inequality and its resources within the city. There are different and diverse kinds of social disparities that are not easy to understand and explore. Geographic information system is an effective research tool for finding hidden social structures and trends in a city. It is hard to accept that Tehran completely and totally bipolarised. To be more precise, there are a lot of well-flourished, well-serviced, regions with good atmospheres and conditions in the southern part of Tehran. On the contrary, there are also many deprived people and regions in the northern part of the city (hidden slums). Changing the scale is essential to detect hotspots such as the Grand Bazaar, which could be an interesting subject for more studies about spatial inequality in Tehran. This challenging place raises the question of what spatial inequality is. Is it related to the context and scale in Tehran or not? The Bazaar is located in the southern part and is very old and in poor physical conditions, with high risks of fire and collapse, but is at the same time one of the most prosperous business and economic centres of Tehran and Iran. The socio-economic inequality within the bazaar is obvious between the rich (businessmen) and the poor (unskilled workers specifically teenagers). On the other hand, there is a sharp and deep land price fault between the Bazaar and its neighbouring areas.

The other problem is that of zoning: the maps that have been used in this research have been produced by the Tehran Municipality based on census tracts, within each of which the data has been aggregated. These census tracts are defined arbitrarily based on some physical barriers or other reasons such as building blocks, transportation networks or official territorial zones.

Last but not least, is the problem of interpolation and extrapolation, and gathering and sampling methods. Spatial analysis techniques could thus distort some facts about (spatial) inequality in Tehran.

13.4.4 Migration

Tehran population growth rate is very high, but there is not a precise household survey with a detailed migration module. Even the Municipality and the central government do not have reliable statistics for Tehran population. Many people come to this megacity permanently, temporarily and daily from all over the country. Tehran population varies in day and night, as a plethora of men, women, students, skilled and unskilled workers active in different economic sectors, specifically in services, come to the city for work, shopping, study, recreation, and various events during the day to Tehran and return to other big and satellite cities near to Tehran at night. The most important of these cities is Karaj, which is part of Tehran metropolitan area with 1,377,450 residents (2006). Also, because of lower land prices (Fig. 13.2) southern parts are more affordable for new-comers and migrants to Tehran (Fig. 13.9) resulting in a higher population growth rate in those parts.

13.4.5 Emerging New West-East Spatial Inequality in Tehran

Studying municipality projects and urban development, and a layover analysis of maps through visual and quantitative exploration, one uncovers a new emerging west-east spatial and social inequality in Tehran, ignored by urban researchers and managers (districts 5 and 22). The western part is more vibrant and developed than the eastern part showcasing more highways, parks, universities, and shopping centres. Spatial demographic analyses provide proof for this new emerging line (Fig. 13.9). The young, well-educated people and the rich prefer the western part to live. Flâneur report (by authors) could be another considerable criterion for testing the hypothesis that these parts are more attractive to these people. It seems that the west, particularly the northwest with its well-developed cultural and recreational spaces and modern shopping centres, is creating a new social class with specific characteristics, namely, educationally well-educated, demographically young, ideologically and politically moderate, originally migrant, socially active and vibrant, and economically unstable. The authors do not intend to label this class, but this new urban social class plays an effective role in urban change, transform and movement in Tehran. The relationship between this new emerging class and the built environment needs more research, but the question is why these people go there? The first reason is that Tehran is surrounded by mountains in north and east parts, these mountains act as physical barriers for further urban development, and the second reason is that the south was not attractive for developers and the real estate market. The only possible area is this western part (as suggested by the Greek planner Constantinos Doxiadis for further urban development in the Tehran Action Plan). On the other hand, this western-northwestern part enjoys a better climate and environmental conditions similar to the north; which may be another reason why people prefer this area. Moreover, land prices are cheaper than elsewhere considering available facilities, urban services and living conditions. It is well connected to other parts of the city through well-developed highways and streets. Recently, Statistics and Information Technology Organisation affiliated to the Tehran Municipality, produced an Atlas of Tehran for different indices such as economic activities, land price, quality of housing, urban services and facilities, air pollution, and population density. All these maps confirm that these young, well-educated people prefer this recently developed northwestern region. Figure 13.10, a cartogram of population density shows and proves this trend very well. This map shows that a large number of people are concentrated in this newly built area, if the data from this map is compared with previous indices. In the southeast and northeast parts, there are dense areas that show high population concentration deserved more study as well.

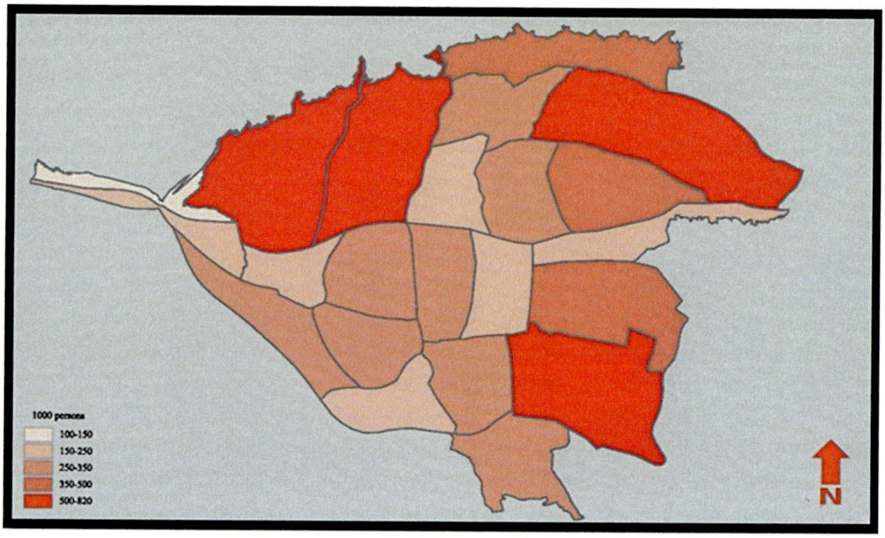

Fig. 13.10 Cartogram of population (Atlas of Tehran Metropolis 2006). Reproduced by permission of Tehran Municipality ICT Organisation

13.5 Conclusion

To conclude, there are four points to highlight:

First, Table 13.1 showed that the production sector in Tehran declined during 1996 to 2006 by 4.27 %. According to Sassen when the production industry starts declining, service industry starts growing in information and communication technology, etc. This table shows, therefore, a consistency with Sassen's statement.

Second, the main idea of the theory of global city by Sassen is that the growth of service industry creates more professional and specialised jobs with higher salaries. Yet in order to support such jobs, low-skill jobs also: these include jobs as cleaners, construction workers, security guards, drivers, etc. In the end, there is the increase from both ends (the rich/top and the poor/bottom) in social structure, resulting in more rich people, more poor people; and the shrinking of the middle classes that used to have a big share in the industrial era as they move either upward or downward the structure. Thus, again, Table 13.1 showing an increase in managerial work by 3.83 %, as well as an increase in unskilled workers of +4.42 % is going in accordance with Sassen's argument.

Third, based on the case of Tehran, there is clearly a social polarisation as explained by Sassen's occupational structure change. Whether or not this polarisation is linked with globalisation is to be tested, yet authors strongly believe that the city is connected to an invisible global network—politically and economically. Forms of spatial inequality in global cities can vary. Using map analysis, Tehran shows a clear socioeconomic north-south divide in the big picture. Explaining such spatial inequality remains a difficult task and there are historical arguments available. The complexity of explanation of spatial inequality is related to different issues; at the macro scale, justifying the events, cultural reasons such as family-tied life style, and the aggregation of data and scale bias can play contradicting roles in this explanation. Furthermore, the new west-east spatial divide which has appeared recently adds more complications to the study of the relationship between social structure and urban landscape, and that authors find worthwhile for interested researchers to look into.

Last but not least, even if we accept that the theory of global city applies to Tehran, we must still consider the fact that social and spatial polarisation in Tehran as a global city was also produced by diverse factors such as different policies specific urban management, land and infrastructure development, and domestic migration. Insufficient spatial policies, carelessness about these policies, and weaknesses of urban management are some of the important factors behind the lack of control in spatial inequality Tehran. On one hand, Iran encountered an eroding and protracted eight-year war and sanctions which cost a lot for the country and isolated it from the world; on the other hand, there were weaknesses of urban management, spatial policy, and mass domestic immigration to Tehran.

13.5.1 Recommendation and Potential Further Research

Using time series maps and mentoring human development indexes by urban researchers can boost socio-spatial knowledge of Tehran, which is now controversial.

Traditional labour market, Grand Bazaar (map8), and business centres, new and old are effective engines for producing socio-spatial polarisation and inequality, and as such they could be bases for interesting socio-spatial studies in Tehran. In another scale the traditional bazaar of Tajrish (map 8) has the same effects for producing spatial inequality around its neighbourhood in northern Tehran.

Tehran is now experiencing a new emerging west-east socio-spatial inequality line as opposed to the old North-South. This could be another area for further urban research. Also, the role of urban sociospatial space in creating this new line can be another reason for local inequality, on top of history, policy and politics: another area in need for further research.

Acknowledgments We would like to thank Professor Costanzo Ranci from the Politecnico di Milano for his patient and constructive comments on this research, and our friend, Yanin Chivakidakarn, Ph.D. Candidate at Politecnico di Milano, for her valuable supports.

References

Amirahmadi H, Kiafar A (1987) Tehran: growth and contradiction. J Plan Educ Res 6(3):167–177

Atlas of Tehran Metropolis (2006) [online]. Available from http://atlas.tehran.ir/. Accessed May 2012

Castells M (1989) The informational city: information technology, economic restructuring and the urban-regional process. Basil Blackwell, Oxford

Ganes HJ (1982) Urban villagers: group and class in the life of italian-Americans. The Free Press (Macmillan Co., Inc.); Updated and Expanded edition (1 June 1982)

Hamnett C (1994) Social polarization in global cities: theory and evidence. Urban Stud 31:401–424

Hamnett C (2001) Social segregation and social polarization. In: Paddison A (ed) Handbook of urban studies. SAGE Publication, London, pp 162–176

Harvey D (1996) Justice, nature and the geography of difference. Blackwell, Oxford

Hayward P, Parent J (2009) Modelling the influence of the modifiable areal unit problem (MAUP) on poverty in Pennsylvania. PA Geogr 47(1)—spring/summer 2009

Madanipour A (1998) Tehran: the making of a Metropolis (world cities series). Academy Press

Madanipour A (1999) City profile, Tehran. Cities 16(1):57–65, printed in Great Britain

Madanipour A (2010) The limits of scientific planning: Doxiadis and the Tehran action plan. Plan Perspect 25(4):485–504

Sassen S (1988) The mobility of labor and capital: a study in international, investment and labor flow. Cambridge University, Cambridge

Sassen S (1991) The global city. Princeton University Press, Princeton, NJ

Sassen S (2001) The global city: New York, London, Tokyo, revised edition (Originally published in 1991), Princeton University Press, Princeton, NJ

Short JR (2004) Black holes and loose connections in a global urban network. Prof Geogr 56:295–302

Tehran Municipality [online]. Available from http://www.tehran.ir. Accessed May 2012

Tehran Municipality Reports on Iran Statistical Centre [online]. Available from http://www.sci.org.ir. Accessed May 2012

Tuan Y-F (1974) Topophilia: a study of environmental perception, attitudes, and values. Prentice Hall

Wacquant L (2008) Logics of urban polarization from below. Polity Press, Urban Outcasts

World Bank (2010) Poor places, thriving people: how the Middle East and North Africa can rise above spatial disparity [online]. Available from http://siteresources.worldbank.org/INTMENA/Resources/OverviewENprintersversionspatialdisparities.pdf. Accessed June 2012

World Bank [online]. Available from http://www.worldbank.org. Accessed May 2012

Chapter 14
Typo-Morphological Analysis of Housing Layout and Density in Tehran

Homeira Shayesteh and Philip Steadman

Abstract This research provides a historical analysis of morphological changes in housing in Tehran, as they have been constrained at each successive period in the city's growth. The research aims to understand this evolution by investigating the relationship between housing typology and urban morphology. A GIS analysis is made of plots, buildings, and blocks in three areas of the city representative of significant stages of development. The analysis focuses on five topics: (1) the question of block size and shape, (2) configurations of plots within blocks, (3) pedestrian and vehicular access, (4) provision of open space around dwellings (ground coverage), (5) natural lighting. The argument of the chapter is that, in the transition from courtyard houses to narrow row houses and later to row apartments, and besides cultural conceptions and the effects of land value, are the constraints imposed by certain 'generic functions', most importantly access (pedestrian or vehicular) and day lighting. This transition is arguably the result of limits imposed on frontage width, which have their origins in the provision of street access to each plot of land in the first phase (from courtyard to row houses); and in the second phase (from row houses to row apartments) the need for more houses and higher densities. The conclusion from the analysis is that although many different cultural, social, physical and environmental factors are involved in the evolution of housing, which should not be treated simplistically, these basic generic functions—such as vehicular access—can dominate all other considerations in defining a new type.

Keywords Housing · Tehran · Typo-morphology · Built form · Plot and block shapes · Access and day-lighting

---

H. Shayesteh (✉) · P. Steadman
Bartlett School of Graduate Studies, UCL, London, UK
e-mail: h.shayesteh@ucl.ac.uk

P. Steadman
e-mail: j.p.steadman@ucl.ac.uk

F.F. Arefian and S.H.I. Moeini (eds.), *Urban Change in Iran*,
The Urban Book Series, DOI 10.1007/978-3-319-26115-7_14

14.1 Introduction

This chapter reviews the growth of the city of Tehran focusing on residential areas. It monitors change in the city over time by looking at the configuration of urban blocks, parcels and streets on one hand, and the built form of residential units on the other. It looks at their interrelationship and the mutual effects that blocks, parcels and housing forms have on each other. The aim is to understand the interaction between the evolution of urban layout and the evolution of residential built forms by providing a quantitative history of the development of residential areas in Tehran in terms of block, parcel and built form sizes, in relation to access and day lighting. GIS analysis is used to establish key quantitative parameters of this evolution: these are block and parcel size, together with access frontage, and the day-lit depths of rooms in buildings. How do morphological constraints, such as access to the street, natural lighting and plot size, affect house types and their physical form over time?

This intermediate scale has not been given the study it deserves in the case of Tehran. Previous research has either been focused on house types in the abstract regardless of their context, that is the urban tissue, describing different types and internal layouts in different regions of Iran (Memarian 1998); or else at the urban scale, investigating the structure of the city (Bertaud 2003; Karimi 1998; Costello 1998). Mirmoghtadaee (2009) studied the process of housing transformation in Iran, but with the emphasis more on life style, although the author also gives a brief descriptive account of the change in housing forms. Azimzadeh (2003) on the other hand studied evolving culture in transforming cities, where he used the internal layout of houses to follow cultural changes in some Iranian cities including Tehran. Madanipour (1998, 1999, 2006, 2011) discussed extensively the concepts and practices employed in the transformation of Tehran, focusing mainly on the urban scale.

The findings of the work suggest that the transformation of courtyards to single-family terrace houses in the first phase of housing development in Tehran was not just an attempt to modernise the city and the result of Westernisation. The desire for vehicular access was a major driving force. For planners at the time, the old house types and street structure were regarded as unable to adapt to modern ways of living that entailed having a car for each family and driving it to the front door. A rectangular grid structure with shallower blocks seemed better fitted for this new style of life. Therefore planners in Tehran limited the depths of blocks to make them two plots deep in every part. This allowed separate vehicular access to all plots. There were also changes in the lengths of blocks, so as to increase the number of narrower plots in a block to obtain improved vehicular access. The second phase in turn was the result of planning controls and the limits they have imposed on ground coverage and density.

14.2 Structure of Tehran's Residential Areas: An Overview

The urban structure of residential areas in Tehran has evolved from deep and relatively irregular blocks to shallower blocks organised in increasingly regular grids. Residential units have also been transformed from courtyard houses to terraced houses with south-facing courts, to flats in multi-storey buildings quite often replacing earlier houses on the same plots of land. In order to quantify these changes and better understand them in physical form, GIS-based measurement techniques are applied to three sample areas in selected districts (boroughs) of Tehran, representative of its successive key stages of development, by reference to the city's growth map. The selected districts are in Boroughs 12, 11 and 2 (Fig. 14.1). Each sample area measures 25 ha (500 × 500 m): each is analysed to find how patterns of access and day lighting of houses have changed. (The GIS maps are from Shayesteh 2013).

During the past century, housing in Iran has undergone substantial changes and this has always been an issue of major concern. In the period between 1925 and 1941 with the development of a grid network of roads through the centres of Iran's old cities, the process of modernisation started in the country and was characterised by a rapid transformation in the structure of those cities, as well as changes in residential districts and housing. This is not to say that the implementation of roads caused the modernisation of the urban fabric; but they were a major part of the process.

Morphological changes in the housing fabric brought about the creation of the terraced house that rapidly replaced the traditional courtyard form (Janipour 2001). These terraces, as the major representative type of contemporary housing, originated in Tehran and spread from there all over the country. From the 1940s onwards

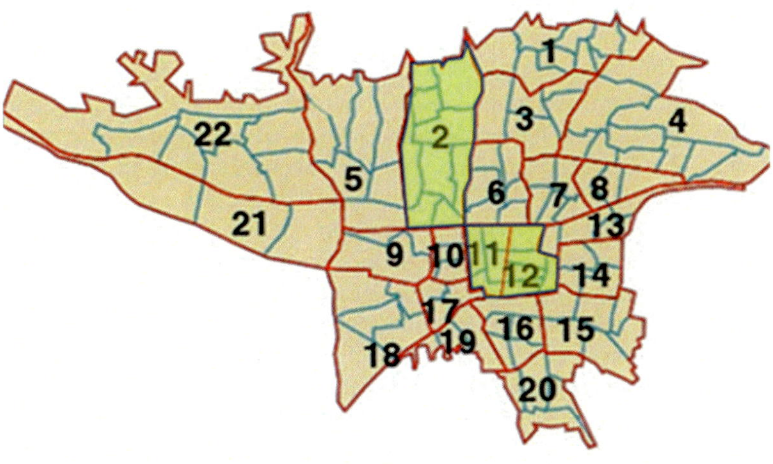

Fig. 14.1 Tehran Boroughs (Shayesteh 2013)

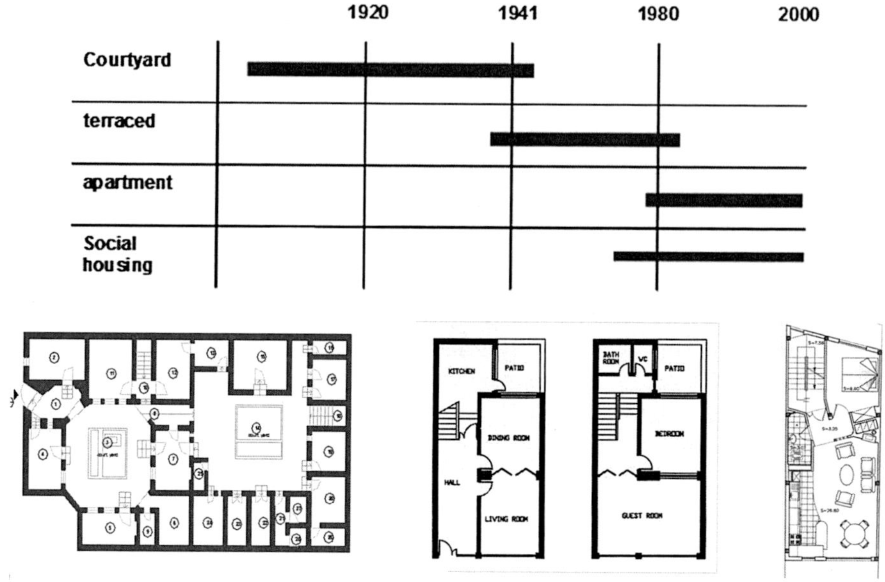

Fig. 14.2 Tehran's housing types, with examples of courtyard house, terraced house, and mid-rise apartment (Shayesteh 2013)

the increase in population and the higher demand for housing caused a huge rise in land values and an associated increase in densities. The low-rise houses (courtyard or terraced) were then replaced by mid-rise 'terraced apartments' when eventually in the 1970s the government allowed people to build multi-storey buildings on plots of land occupied by houses, where this had not previously been permitted, as well as in new developments (Fig. 14.2).

Single-family old houses built from brick and wood, mainly on one or two storeys, were thus replaced by multi-storey apartments. A few high-rise apartments were also built for government employees and lower and middle class families. All these resulted in there being more construction in areas of the city that were already densely developed. The whole process was financed by homeowners' individual investments. The lack of an overall plan for street facades and also the lack of coordination resulted in a very chaotic skyline (Fig. 14.3).

In addition to the need for vehicle access to the streets, modern techniques of construction also had a major influence on the morphology of the city and the urban fabric. This phenomenon is important because it has occurred not just in the new developments and suburban areas, but has been accompanied by an enormous amount of demolition in the historical areas.

Borough 12 embraces the nucleus of the city, its original parts, and the bazaar. It contains a great number of traditional courtyards that were then integral to the city. Due to the original system of narrow and 'organic' pedestrian routes in this area some of these houses still do not have direct vehicular access. For those that do

Fig. 14.3 Redevelopment at different times leaves the skyline of Tehran's streets fragmented: example of a street in Borough 2 (Shayesteh 2013)

have vehicular access, this is often very narrow. Courtyard dimensions vary considerably; however, the majority of plots have dimensions between 3 and 20 m. The average area of plots of land in Borough 12 is around 190 m^2. Looking at the size of blocks in the 500 × 500 m sample area shows that block sizes vary between 330 and 11,315 m^2 with the mean being 4420 m^2. The absolute minimum size of block is in fact a case where a single plot has street access on all sides. The population of this borough is 248,048 according to data from the Statistical Centre of Iran web publications in 2006 (http://www.citypopulation.de/Iran-Tehran.html).

Borough 11 contains later phases of development, which constitute a 'transitional stage' from an organic traditional city with an irregular street network and courtyard houses to a modern city with a rectangular grid and similar row houses. At this stage a transitional type of house emerged where the form changed from a completely enclosed courtyard, to the plot being developed in two structures at front and back with an open space in between. Examples of this typology can be seen in Borough 11. The population of this borough was 275,241 in 2006 (http://www.citypopulation.de/Iran-Tehran.html).

Borough 2, the largest of the three, contains developments that are more recent. Within this borough, the southern parts are older, and as the city was extended to the north, residential areas were developed in this direction. As is clear from the map the majority of the urban fabric in this borough is modern and planned on a rectangular grid. The population of this borough in 2006 was 608,814 (http://www.citypopulation.de/Iran-Tehran.html).

14.3 The Question of Block Size and Shape

Block statistics in the three sample areas within each borough are presented below. A ‘block’ means a piece of land bordered on all sides by streets or other vehicular routes. Borough 12 and 11 sample areas consist of a variation of different-sized blocks whereas the Borough 2 sample area consists of blocks of a standard shape and size. Borough 12 has more blocks in the smaller size ranges, while in the Borough 11 sample area we see most variations in size; indeed we find the largest blocks in this borough. What does this all mean from a morphological point of view?

Obviously, the Borough 2 sample area speaks of planning attempts to unify the urban blocks in terms of size and shape according to modern town planning principles and is the result of a clear pattern of land division. This area is an example of standardised planning practice for a rapidly changing city needing more houses.

The Borough 12 sample area, however, is an example of a spontaneous configuration of blocks with gradual additions to the plots. Here the size of blocks has been more deliberately managed towards the small and medium size ranges, perhaps by locals themselves. The scale is obviously arranged to suit pedestrian movement. The smaller the urban blocks, the more accessible they are. The reason is that smaller blocks have more road space per unit of built area and less diversion of routes. It is interesting how this self-management resulted in the formation of a more accessible block structure. This of course responds to pedestrian accessibility and is because most movements were on foot when the blocks were formed. Larger blocks are more suitable for cars.

The Borough 11 sample area, the intermediate phase between spontaneous and designed town planning, shows some interesting results. It not only has the greatest variation in block size but also the largest blocks amongst the three sample areas. This shows the level of experimentation that planners made over a short period.

In addition, a comparison of the three sample areas shows that Borough 11 has the lowest number of blocks indicating the lowest level of accessibility. The number of blocks in the selected area of Borough 11 is half of that in Borough 12. Table 14.1 gives the number of blocks within each sample area. It is important to point out that blocks going over the boundaries of the 500 × 500 m frames have been excluded from the analysis. Figure 14.4 also illustrates block shapes and sizes in the three boroughs. From the thematic maps it is clear that block sizes in the range 3–6000 m^2 occur with the highest frequency.

Let us look at Table 14.1 in more detail. The area of each block was measured automatically in GIS. However, the dimensions of each block presented a more

Table 14.1 Block analysis for sample 500 × 500 m areas (Shayesteh 2013)

	Borough 12	Borough 11	Borough 2
Number (blocks)	38	19	23
Mean area (m^2)	4420	6213	5502
Modal area (m^2)	3–6000	3–6000	3–6000
Mean dimensions (m)	D1: 50 D2: 95.5	D1: 63 D2: 103	D1: 56 D2: 99
Modal dimensions (m)	D1: 30–45 D2: 55–118	D1: 42–94 D2: 108–145	D1: 55–57 D2: 98–99
Mean ratio	1:1.9	1:1.7	1:1.6
Modal ratio	1:1.4–1:1.8	1:1–1:1.6	1:1.7–1:1.8

Fig. 14.4 Thematic map of block area in sample areas. (From top to bottom, Borough 12, Borough 11 and Borough 2) (Shayesteh 2013)

complex problem, since the blocks were not all rectangular. The decision was made to approximate the shapes of blocks to the closest rectangular form. Moreover, the smallest dimension of the block in each case is in fact an approximation of the shortest side (D1) of each block. This measurement was made manually. The longer dimension (D2) was the result of dividing the area of the block by the measurement of the shorter side. Average dimensions for the shorter and longer sides of each block were then calculated to give an 'average block' for each borough. However, as is evident from the map, the range of block shapes and dimensions is quite wide in Boroughs 12 and 11, unlike Borough 2 where blocks are all of the same size and shape. The mean values must therefore be seen in relation to these distributions.

The modal values for block dimensions (the most frequent interval in the distribution) were also calculated. The 'mean ratio' and 'modal ratio' give the ratio of D1 to D2 in the relevant rows above. They give an indication of the typical shapes of block in each borough.

Over time, there is a trend towards shallower blocks. It is clear from the maps of the sample areas that in the older parts, some blocks were more than two plots deep. In more recent developments, where it is evident that blocks are the result of planning practice, it is extremely rare to find a block more than two plots in depth. The demand for vehicular access is certainly the reason behind this.

14.4 Configuration of Plots Within Blocks

It is clear that in the modern areas of the city, where blocks are the result of town planning decisions, areas are developed by subdividing the land in most cases into simple rectangular plots.

Examples can be found in Borough 2 and parts of Borough 11. So following our earlier argument, we can say that the issue of the configuration of plots within urban blocks is related directly to land subdivision. This is not the case in the older parts of the city, where it seems that the development of a block was the result of a gradual, spontaneous process of agglomeration of plots and houses over longer periods. Examples of such blocks can be seen in the sample area of Borough 12.

Parcel areas and building footprint areas were measured for all parcels within each sample area. Thematic maps of parcel size and building footprint area were also prepared, allowing initial comparisons for the three sample areas. The graph of Figs. 14.5 and 14.6 show the distribution of plot size in the three 500 × 500 m areas. We see a similar pattern in the older Boroughs 12 and 11 in all of the plot size categories. The sample areas in Boroughs 12 and 11 have the highest proportion of plots (50 and 52 %) in the 100–200 m^2 range. In addition, they have the majority of plots in the first three smallest categories. Plots of the size 300–400 m^2 in Boroughs 12 and 11 show a lower percentage (still considerable), whereas plots with areas greater than 400 m^2 in both boroughs have totals of 4 and 5 %. Regarding the sample area in Borough 2, the proportion of 200–300 m^2 plots is considerably higher than any other plot category, with a value of 73 %.

This indicates how homogeneous this most extensively and uniformly planned area in Borough 2 is in terms of plot size, compared to other boroughs. The percentages of small plot size categories are particularly low, with values of just one percent for both 50–100 and 100–200 m^2 plots. In addition, plots of the size of 500–5000 m^2 also have a high proportion of 17 %. This is the result of amalgamations of standard-sized plots, as well as wider plots created in the initial land subdivision.

The graph of Fig. 14.7 presents the distributions of building footprint size in the sample areas of the three boroughs.

The graph again shows similarities in the distribution of footprint area in the older Boroughs 12 and 11. Both sample areas in Boroughs 12 and 11 have the highest proportion (38 and 42 %) in the 30–80 m^2 range. They both have their largest proportion of building footprints in the first and third size categories (30–80 and 100–140 m^2). Regarding the sample area in Borough 2, the percentage of 140–200 m^2 buildings is much higher than any other plot category, with a value of 70 %.

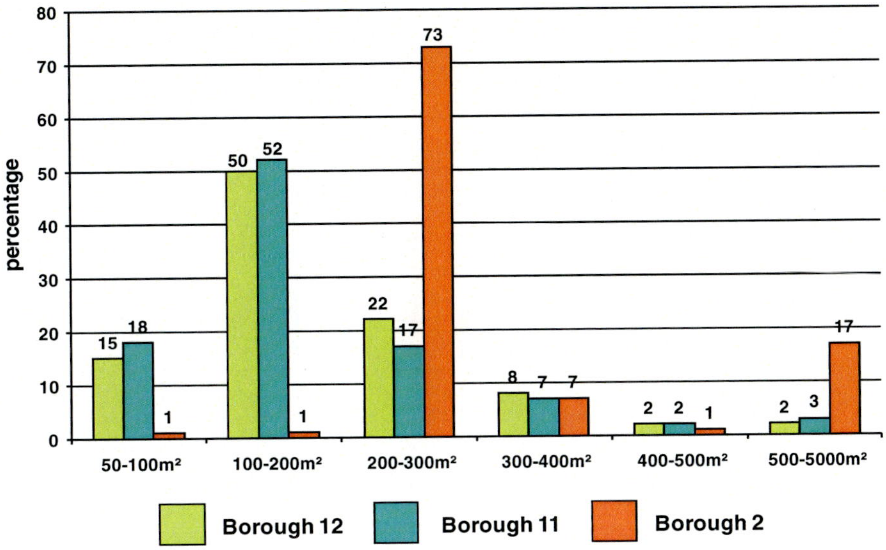

Fig. 14.5 Parcel distribution in 500 × 500 m sample areas (Shayesteh 2013)

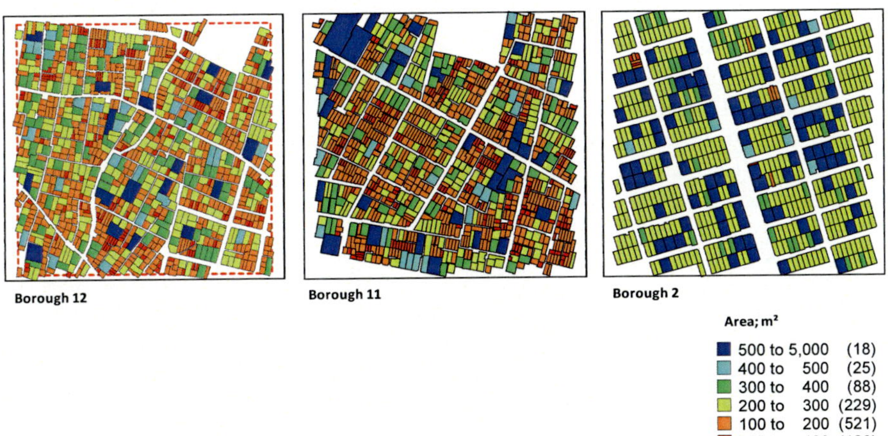

Fig. 14.6 Parcel size (m^2) in clockwise from top to left to bottom right, Borough 12, Borough 11 and Borough 2 (Shayesteh 2013)

This, once again confirms how homogeneous this planned area is in terms of both plot size and footprint area, compared to the other boroughs. In addition, building footprints in the band 300–3000 m^2 also have a high percentage of 18 % followed by the category of 200–300 m^2 with 10 %.

The maps in Fig. 14.8 show the analysis of footprint size. The map of Borough 12 reveals a relatively even distribution throughout the sample area. There are,

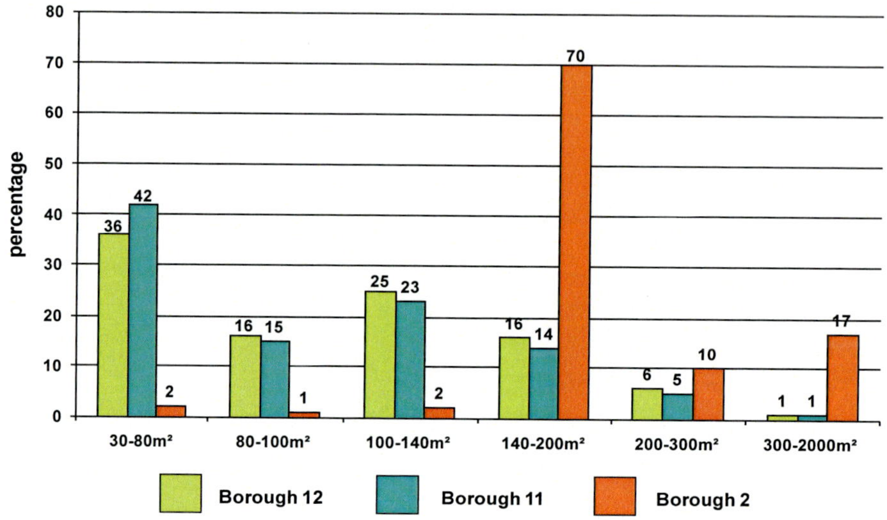

Fig. 14.7 Footprint size distribution (m^2) in 500 × 500 sample area

however, some locations with a greater concentration of smaller footprints (shown by black circles). This is an indication of the more gradual growth of the area with a less uniform type of planning. There are also very few larger footprints (300–2000 m^2).

The footprint analysis in Borough 11 demonstrates a wide variation of different footprint size groups. The number of larger footprints is still small. There is also a shift towards greater zoning of similar footprint size groups. This indicates that

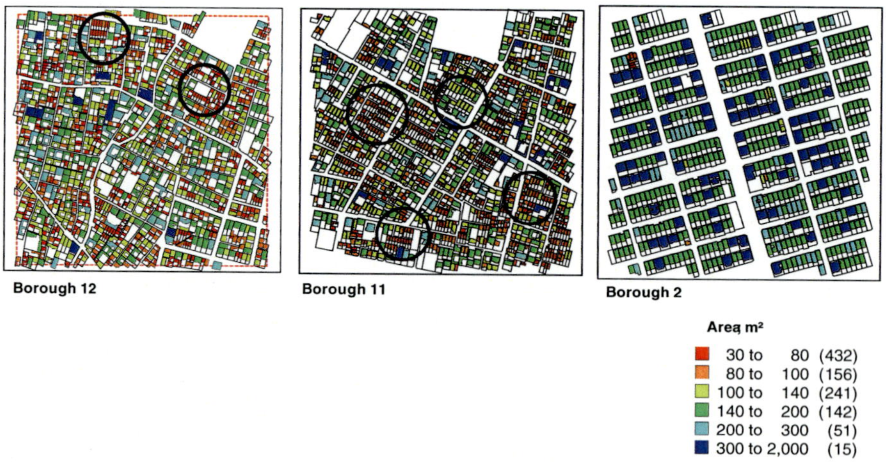

Fig. 14.8 Footprint size (m^2) in clockwise from top left to bottom right, Borough 12, Borough 11 and Borough 2 (Shayesteh 2013)

town planning practice has been more instrumental in the formation of this area, and that the area was developed in a shorter time span than the Borough 12 sample area.

Finally, the analysis of the Borough 2 area shows the greatest standardisation in the size and shape of building footprints. Here there is a shift towards larger footprint categories, and little variation in size. This is a pure example of planned subdivision. The entire area is divided into similar size blocks and similar size plots of land. Some smaller plots are the result of bisecting the original plots. In addition, some larger plots are the result of amalgamations of smaller plots. It should be noted here that a planning regulation that has been enforced in Tehran since 1955 requires the buildings to be positioned only on the north sides of plots. The maximum extent of ground coverage is also specified. This regulation has had significant implications on the overall built form of the city.

14.5 Pedestrian and Vehicular Access

The front edge of each plot that touches the edge of the street or pavement and maintains the access to the plot (i.e. the frontage length) was also measured for plots in all three sample areas. This measurement allows an investigation of the relationship between typology and frontage. Some assumptions were made in order to define which side of a plot is the frontage in special cases, which were relatively numerous. In cases where a plot has more than one exposed side facing the street, for example a corner plot (i.e. a plot also accessible from the side), the side facing the wider or major road is counted as the frontage.

The 500 × 500 m sample areas in each borough are shown in Fig. 14.9, with the access frontages coloured. The analysis in the Borough 12 sample area shows that

Fig. 14.9 Frontage width (m) in clockwise from top to left to bottom right: Borough 12, Borough 11 and Borough 2 (Shayesteh 2013)

plots along the main streets have in general wider frontages. Going to smaller streets and alleys, the frontage width becomes narrower. The second finding of this analysis is that there is a wide variation in frontage widths, and there are many cases where adjacent plots have different frontage widths. This is another indication of a gradually evolved urban fabric in this part of the city. A further finding is the large number of interior sites: plots with no street access frontage that normally have pedestrian access from other plots or via little private covered alleys. These are marked in the map by small red dots.

In Borough 11 the analysis shows that, as in Borough 12, the wider frontages belong to major roads. However, the differentiation of frontage widths between adjacent plots is less marked than in Borough 12, and there are also fewer interior sites. These show the consequences of some early town planning interventions in the area.

Borough 2 as a typical planned development has more uniformity in widths of frontages. North-south major streets have wider frontages than east-west streets. The standard plots result in similar frontage widths. The majority of plots also have wider frontages than in Boroughs 12 and 11. The bar chart of Fig. 14.10 shows the distribution of frontage width in the three boroughs, and indicates that in Boroughs 12 and 11 there is greater variety of width categories, while in Borough 2 the smallest categories are almost completely absent.

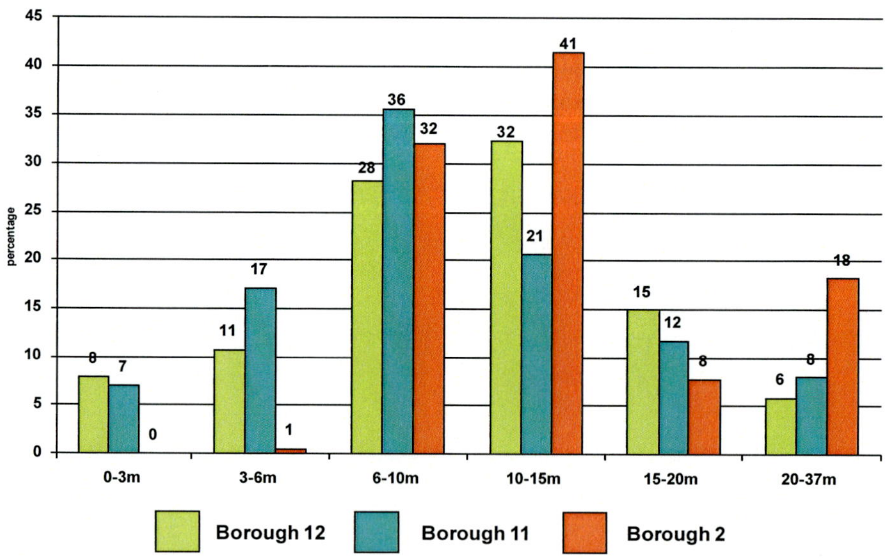

Fig. 14.10 Distribution of frontage width (m) in 500 × 500 m sample areas (Shayesteh 2013)

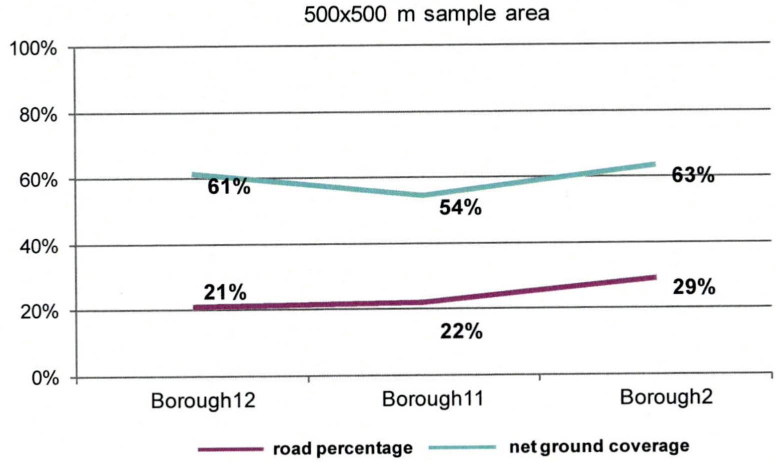

Fig. 14.11 Relationship between percentage road area and net ground coverage (Shayesteh 2013)

14.6 Provision of Open Space Around Dwellings

The provision of open space around buildings was investigated measuring variables including the percentage of land area devoted to roads, gross ground coverage by buildings as a percentage of the sample area, and net ground coverage as a percentage of total plot area. (Ground coverage is also sometimes referred to as 'building ratio'.) The ground coverage in Borough 12 was 61 %. The modern legislation on ground coverage referred to earlier sets a limit of 60 %: it seems possible that this figure was derived from measurement of the historical situation in Borough 12.

According to the graph of Fig. 14.11, the percentage area of roads showed a small increase in Borough 11 compared to Borough 12, from 21 to 22 %. Later in the Borough 2 area however, there is a significant increase to 29 %, which is what one would expect from greater road widths for vehicular traffic.

14.7 Natural Lighting

Access to natural lighting is a very important measure of the quality of buildings in general and houses in particular. The angle of the sun, the width over which the building is exposed to natural light and the depths of rooms as well as the dimensions of windows (and other openings), are all crucial in this respect.

In this study the sizes and numbers of windows have not, however, been taken into account. The emphasis instead is given to the width over which the buildings are exposed to the exterior (day-lit frontage width) and the depths of the rooms in plan (day-lit depth). Day-lit frontage width was measured for buildings in each 500 × 500 m sample area. It is important to differentiate between access frontage width and day-lit frontage width, although these can, in some cases, be the same. Access frontage is the measurement of the width over which the plot of land is accessible by either pedestrians or vehicles. Access frontage width deals with the plot itself; however, day-lit frontage width measures the length over which a building can have openings either to its own private open space or to public open spaces and roads.

The maps of Fig. 14.12 show an example in Borough 12 explaining how access frontage and day-lit frontage width are calculated. For the plots in the centre of the map (A, B and C), the day-lit frontage is shown in bright red and the access frontage is shown in dark blue (access from a small alley in plots A and B). As seen in the map, in plot C the side of the building adjacent to the alley is considered both as day-lit frontage width and access frontage width. This is not the case in plots A and B, which have very small access frontages from the alley, and obtain their daylight from the courtyards.

Day-lit depth is calculated by dividing the total area of the building on the ground floor by the day-lit frontage width. This is done because the calculation of the actual day-lit depth for each individual building was complicated, for two main reasons. The first reason was that interior plans were not available, and the second was that there were many cases where one room had multiple aspects. The decision to make the calculation this way means that the results are somewhat less precise, but for the purpose of this study the figures are adequate. In summary, the calculation gives the day-lit depth for a strip within a rectangular built form of the same area as the actual building, lit from only one side, and with the same dimension as the day-lit frontage width.

The graph of Fig. 14.13 shows the changes in mean values for day-lit width and day-lit depth across the sample areas in the three boroughs. It is clear that Borough 11 has narrower day-lit frontage but more day-lit depth in comparison to Borough 12. This shows that the process of providing direct vehicular access to each plot, as the main agenda in this phase of transformation, has resulted in shorter day-lit frontages and deeper rooms.

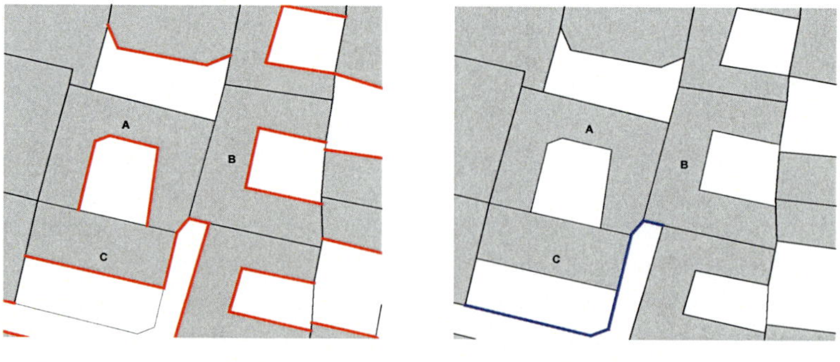

Fig. 14.12 *Left* day-lit frontage, *Right* access frontage in Borough 12 (Shayesteh 2013)

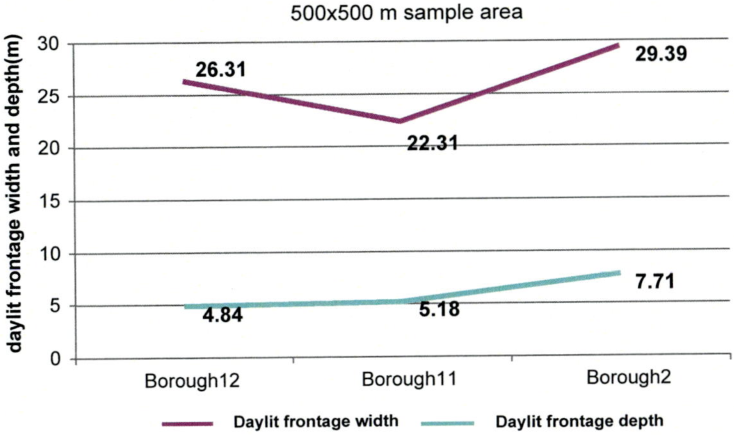

Fig. 14.13 Day-lit frontage width and day-lit depth across the three boroughs (Shayesteh 2013)

However, in Borough 2 a slightly different pattern is evident. Despite a significant increase in day-lit frontage width, rooms become notably deeper. This is to do with increase in density and ground coverage in house types. The increase in frontage width is the result of generally larger plots of lands in this borough.

14.8 Findings and Discussion

The research aimed to understand the evolution of housing in Tehran by investigating the relationship between housing typology and urban morphology, through an analysis of the physical characteristics of plots, buildings, and blocks in three areas of the city representative of its successive stages of urban development.

It was found that the urban parameters of block and plot size do make certain types of built form more likely than others. It was also found that the impact of these urban parameters is mediated by building regulations—effects that could be the subject of a future study.

Table 14.2 shows some parameters of urban space and development including ground coverage and the proportion of surface devoted to roads, together with parameters of built form and parcel size and shape, including access frontage width, day-lit and court depth, parcel area, and building footprint area. According to the GIS analysis, the total numbers of buildings in the 500 × 500 m sample areas were 1114, 1037 and 546 in Boroughs 12, 11 and 2 respectively. The first thing that stands out from the table is the reduction in the number of plots (variable B) and subsequent fall of the number of buildings across the sample areas (variable J). Plots almost halved in number in the Borough 2 area compared with the older boroughs. This is an indication of the plots being larger in area. According to the

Table 14.2 Summary of comparisons of dimensions between sample areas in the three boroughs

Variable	Details	Borough 12	Borough 11	Borough 2
A	Sample area	500 × 500 m 25 ha/62 acres	500 × 500 m 25 ha/62 acres	500 × 500 m 25 ha/62 acres
B	Number of plots	1041	1031	556
E1	Mean (average) plot size (m^2)	190.41	191.21	323.11
E2	Modal plot size (m^2)	100–200	100–200	200–300
F	Area covered by plots (m^2)	198,227	197,138	179,650
I	Roads as % of sample area (%)	21	22	29
J	Number of buildings	1114	1037	546
M	Mean (average) footprint (m^2)	108.83	104.52	210.10
P	Net ground coverage (as % of plots area) (%)	61	54	63
S1	Access frontage (mean) (m)	10.75	10.20	13.67
S2	Access frontage (modal) (m)	10–11	6–7	9–11
V	Plot depth (mean) (m)	16.53	17.6	25.5
Y	Day-lit frontage width (mean) (m)	26.31	22.31	29.39
A2	Day-lit depth (mean) (m)	4.84	5.18	7.71
A3	Day-lit depth (modal) (m)	3–5	4–6	8–10
A4	Court depth (mean) (m)	9.0	9.3	10.6
A5	Court depth (modal) (m)	7–9	8–9	9–11

table the mean plot size (variable E) showed a significant rise in Borough 2 in comparison to Boroughs 12 and 11. The mean plot size was almost 190 m^2 in Boroughs 12 and 11, with an increase to almost 323 m^2 in Borough 2. The value for court depth, measured by taking the square root of (plot size—footprint size), also increased through time which is the result of overall longer rectangular standard plots. Changes in day-lit depth affected some other properties of parcels including court depth. The two variables are systematically related.

14.9 Conclusion: Vehicular Access and Density as Underlying Forces in Housing Evolution

The interaction between built form and urban form is an important field of study from multiple points of view, including environmental performance at the building and urban scales, the organisation of space to support specific lifestyles, and the fulfilment of requirements concerning access and transportation.

The pattern, size, shape, and arrangement of buildings, plots, and street blocks can produce very different results in terms of built form and density (Steadman

1994, 2003). It seems that certain combinations of plot dimensions and arrangements lend themselves better to particular kinds of house types. The question is then: *do differently sized plots create or generate different types of houses?*

Taking an overview of the whole process, the transformation of courtyards to single-family terrace houses in the first phase of housing transformation in Tehran was not just an attempt to modernise the city and the result of Westernisation. It was the desire for vehicular access which was the main driving force. As was shown in the analysis, ground coverage did not increase during this phase of Tehran's transformation, because that issue was not on the agenda for decision makers. In fact, the net ground coverage decreased from 61 % in Borough 12 to 54 % in Borough 11 (variable P, Table 14.2). This shows that for planners at the time, the old house types and street structure were regarded as unable to adapt to modern ways of living that entailed having a car for each family and driving it to the front door. They might have thought that a rectangular grid structure would better support this way of living. They therefore limited the dimensions of blocks, thus restricting the number of plots in a block to maintain vehicular access to each plot. The question of what criterion directly governed the *lengths* of blocks remains a difficult one to answer. What is more relevant is that planners in Tehran limited the depths of blocks to make them two plots deep in every part. This made possible separate vehicular access to all plots. By comparison, the older blocks where there are some interior plots are much deeper.

This research has attempted to quantify the historical development of residential areas in Tehran in relation to their blocks, parcels and built form dimensions. In this process, blocks became shallower over time. There was a trend towards narrower plots, which meant that full courtyard forms for houses could no longer be accommodated. Rooms in houses also became deeper over time and the quality of day lighting deteriorated. The block structure of Tehran in recent developments has been affected by building regulations.

Hence, the transformation process analysed in this chapter shows that although many different cultural, social, physical and environmental variables and factors are involved in the evolution of housing, there are also some basic generic functions—in particular vehicular access—that can dominate over other factors in defining a new type.

References

Azimzadeh M (2003) Evolving urban culture in transforming cities, architectural and urban design in a fluid context. Ph.D., Department of Urban Design and Planning, Chalmers University of Technology, Göteborg, Sweden

Bertaud A (2003) Tehran spatial structure: constraints and opportunities for future development. Ministry of Housing and Urban Development, Islamic Republic of Iran, National Land and Housing Organization, National Housing Committee

Costello VF (1998) The morphology of Tehran: A preliminary study. Built Environ 24(4):201–216

Janipour B (2001) Periods of changes in the housing architecture in Tehran (under the rule of Pahlavi dynasty). Ph.D., University of Tehran, Faculty of Fine Arts

Karimi K (1998) Continuity and change in old cities; an analytical investigation of the spatial structure in Iranian and english historic cities before and after modernisation. Thesis (Ph.D.). The Bartlett School of Graduate Studies, University College London

Madanipour A (1998) Tehran: the making of a metropolis. John Wiley & Sons Ltd, Chichester

Madanipour A (1999) City profile Tehran. Cities 16(1):57–65

Madanipour A (2006) Urban planning and development in Tehran. Cities 23(6):433–438

Madanipour A (2011) Sustainable development, urban form and megacity governance and planning in Tehran. In: Sorensen A, Okata J (ed) Megacities: urban form, governance, and sustainability, pp 67–92

Memarian G (1998) Persian house typology. Thesis (Ph.D.). The University of Manchester

Mirmoghtadaee M (2009) Process of housing transformation in Iran. J Constr Developing Countries 14(1):69–80

Shayesteh H (2013) Typo-morphological approach to housing transformation in Tehran. Thesis (Ph.D.). Bartlett School of Graduate Studies, UCL

Steadman P (1994) Built forms and building types: some speculations. Environ Plan 21:S7–S30

Steadman P (2003) How day-lighting constrains access. In: Space syntax 4th international symposium, 1st ed. Space Syntax Laboratory, London, pp 05.1–05.18

Part V
Shaping Iranian Cities (Zooming In): Urban Projects and Programmes

Chapter 15
The Effects of Iran's First Baby Boomers (1976–1986) on the Housing Economy of Iran and the Government Policies to Deal with Its Resulting Issues

Seavash Jamali and Manoochehr Dadashzadeh

Abstract Iran's first baby boomers were born between 1976–1986. The country's annual population growth rate during this period is said to be 3.9 %, which means 16 million people were born within ten years. Population in the first and final year was recorded as 33 and 49 million respectively. Nowadays, there are roughly 16 million 25–35 s in need of residential units, causing extra demands for living places in the housing market and leading to high inflation in the housing economy, and house prices are bound to significantly soar. In this chapter we discuss the impacts of this growing demand for the housing economy, and show that the government solutions for this problem, such as new towns and Mehr Housing Project (Maskan-e-Mehr) are not successful owing to the fact that these solutions are just emphasising on building houses, ignoring other vital aspects of housing projects (such as services, creating job opportunities and the like). Furthermore, we provide some remedial suggestions for these ongoing policies.

Keywords Housing economy · Demand and supply of residential units · Government housing policies · Mehr housing project

15.1 Introduction

Housing is of vital importance to every individual (Aronovici and McCaknont 2010). After rapid population growth and urban development, one of the most important problems especially facing developing countries is housing. Iran has

S. Jamali (✉) · M. Dadashzadeh
Science and Research Branch, Azad University, Tehran, Iran
e-mail: seavash@gmail.com

M. Dadashzadeh
e-mail: Manoonchehr.dadashzadeh@gmail.com

F.F. Arefian and S.H.I. Moeini (eds.), *Urban Change in Iran*,
The Urban Book Series, DOI 10.1007/978-3-319-26115-7_15

faced this in recent decades, but it is exacerbated today by another problem according to the Centre of Iranian Statistics: annual population growth rate in Iran between 1956 and 1966 was 3.1 %, so 6.9 million people were born during that decade. Between 1966 and 1976 the rate was 2.7 %, resulting in 7.9 million new births, and between 1976 and 1986, 3.9 % with 16 million people born in that period. Iran's social policies during the Iraq-Iran war resulted in a baby boom (Wikia 2012) initially starting with the 1979 Islamic Revolution, during which families were strongly encouraged to contribute to a baby boom (National American daily newspaper 2012).

The population of 33 million in 1976 rose to 49 million in 1986: in other words about half of the population of 1976 (almost 16 million) were added to the country's population between 1976 and 1986. Studies carried out in the countries with similar experiences have attempted to investigate the effects of baby boomers on the housing market (Myers and Ryu 2007). Myer (2007) explains how the giant baby boom generation (in the US) has been a dominant factor shaping the housing market. In his study he focuses on population in the age range between 24 and 64. It means that adults older than 24 are most likely to own homes. Nowadays there are 16 million people within the 25–35 bracket, and by the same token, this population is most likely to own homes in Iran. Analysing implications of demographic trends driven by human aging can help planners envision the changes ahead (Masnick 2002). After entering into housing market, people of this age range will generate a huge demand for living places in the housing market leading to high inflation in the housing economy and the price of houses and apartments will significantly increase. What made the Iranian situation worse was the population growth between 1986 and 1996, with its 10.5 million new births, which, in the absence of proper planning can lead to an exacerbated situation. According to Hopkins and Zapata (2007) 'as planners we want to work with constituencies to engage and shape futures, not merely stumble upon these futures as they emerge. To shape futures, we must imagine them in advance and understand how they might emerge' (Hopkins and Zapata 2007). Demographic changes in Iran between 1956 and 1996 are shown in Table 15.1.

Main questions here are:

- What are the effects of entering baby boomers on the housing economy?
- What are government policies to deal with problems of housing?
- What are the deficiencies of these policies and how to correct those?

Table 15.1 The increasing rate of population in Iran between 1956–1996 (Iranian Statistics Centre, n.d.)

Years	1956–1966	1966–1976	1976–1986	1986–1996
Yearly growth rate (%)	3.1	2.7	3.9	1.9
Increasing population	6,900,000	7,900,000	16,000,000	10,500,000

Table 15.2 Population, households, housing units and households changes in Iran during 1966–2011 (Iranian Statistics Centre, n.d.)

Year	1966	1976	1986	1996	2006	2011
Population	25,788,722	33,708,744	49,445,010	60,055,488	70,495,782	75,149,669
Households	5,029,320	6,709,068	9,548,888	12,280,539	17,464,062	21,185,647
Housing units	3,898,719	5,305,538	8,217,375	10,770,112	15,859,926	19,954,708
Household size (people in each family)	5.1	5.0	5.1	4.8	4	3.5

15.2 Housing Condition in Iran

Table 15.2 shows that between 1996 and 2006 there was a growing trend in real estate in Iran. For instance, 5,183,523 households and 5,089,814 housing units were added during 1996–2006 (in ten years). Also, 3,721,585 households and 4,094,782 housing units were added during 2006–2011 (in five years). The data show a sharp change in the household size between 2006 and 2011, and there will be even a sharper change in household size between 2011 and 2016 because, according to statistics, between 2006 and 2016 about eight million housing units will be needed whilst only 4,094,782 of these units will have been built, so during 2011–2016 there will be a demand for about four million more housing units in Iran.

Based on the age pyramid of Iran (2011) 7,525,894 people are between 30–35 and 9,044,823 between 25–30 and these people are highly likely to enter the housing market. Thus in the next five years there will be a huge demand in the housing market of Iran (due to the fact that the average marriage age in Iran is 25, and married people usually enter the housing market within two or three years after their marriage).

15.3 The Influence of Baby Boomers Entering into the Housing Market

The main question here is whether baby-boomers are able to create effective demand? Demand is defined as the quantity that consumers are able and willing to purchase at each conceivable price (Wikia 2012). It means effective demand is based on the desire and also ability of consumers to buy something. Most people of the young generation in Iran cannot afford housing prices, and therefore unable to effectively enter into the housing market. Thus, they will enter rental market instead, making this market so profitable that upper classes are encouraged to buy houses to let, which in turn results in an effective demand in housing market.

15.4 Government Housing Policies

According to article 11 of ICESCR (International Covenant on Economic, Social and Cultural Rights), signed by Iran in 1976, every person has the right to an adequate standard of living, which includes the right to adequate housing (International Covenant on Economic, Social and Cultural Rights 2012). Also, according to article 31 of the Constitution of the Islamic Republic of Iran, it is the right of every Iranian individual and family to possess housing commensurate with their needs. According to this article the government is obliged to make land available for the implementation of this article, prioritising those in greatest need, particularly the rural people and the labourers.

During various periods, the Iranian government took steps to deal with the housing problem such as initiating the Mehr Housing Project (Maskan-e-Mehr), establishing new towns, and so on. What follows is a scrutiny of the effects of the New Towns policies and the Mehr Housing Project to demonstrate their deficiencies.

15.5 Planning New Towns

New town planning has been based on political and military objectives, particularly in providing housing facilities and shelter for workers in the oil industry during the pre-revolution period (prior to 1979). Post 1979, they were constructed with the aim of decentralisation, attracting surplus populations of large cities, offering employment opportunities and providing affordable housing and also relocating some industrial establishments away from large cities (Ziari and Gharakhlou 2008). Below is a more detailed description of these goals:

- Controlling the rapid and disorderly growth of large cities by limiting the size of their populations.
- Absorbing 6 million surplus population of big cities until 2021, by offering them new employment opportunities in surrounding New Towns. These were not to be planned as dormitory communities: but to offer people a variety of employment opportunities. The aim was to overcome urban housing problems and to provide low-cost, affordable housing in new settlements to deal with high cost of living in large cities.
- Relocating some inappropriate industrial establishments from the big cities to the New Towns.
- Preventing the formation of informal settlements in the periphery of the large cities (Ziari 2006).

But now, after two decades of planning and building them, a close scrutiny of the condition of these new towns reveals that they have been unsuccessful in achieving their goals. Table 15.3 shows a list of new towns and their populations.

Table 15.3 The list of Iran's recent new towns of Iran [Combined data from Ziari (2008) and Iranian Statistics Centre]

Parent city	Name of new town	Distance form parent	Area (ha)	Established	Predicted population	Present population (2009)	Ratio of the present population to the projected population (%)
Tehran	Hashtgerd	65	4000	1990	500,000	17,000	13.3
	Andisheh	30	1100	1991	132,000	76,122	
	Pardis	25	2000	1991	200,000	30,000	
	Parand	40	1467	1990	150,000	8000	
Isfahan	Baharestan	15	3000	1989	500,000	60,000	10.8
	Poladshahr	25	7000	1989	500,000	59,616	
	Majlesi	65	3000	1989	140,000	4250	
Mashhad	Golbahar	40	4000	1990	400,000	7387	2.1
	Binalod	45	4000	1991	113,000	3651	
Shiraz	Sadra	15	2040	1991	200,000	17,984	8
Tabriz	Sahand	20	3148	1991	100,000	15,992	15.9
Ahvaz	Ramin	35	1000	1990	120,000	–	
	Shirinshahr	35	1000	2002	110,000	–	
Total					3,165,000	30,002	0.94

As it shows, in 2009 only about one percent of projected population of these new towns were actually living in them, despite the fact that most of these cities provided affordable houses. There were a lot of reasons why these new towns were unsuccessful:

- A lack of proper location (far from larger cities and lacking proper transport links).
- A lack of job opportunities and dependence of the residents on parent cities.
- A lack of identity in new towns as most residents were from different towns and cities.
- Social conflicts, due to contrasting cultural backgrounds.
- A lack of public services in these towns.

15.6 The Mehr Housing Project

Aiming at constructing housing units for the vulnerable and low income population, the Mehr Housing Project provided another opportunity for the Government to solve the housing problem of labourers (Islam Online 2012). Paragraph 'D', Note 6 of the Government Budget for the year 2007–8 has put a strong emphasis on the provision of adequate housing for all citizens, especially low-income ones, and has proposed several mechanisms for achieving the determined objectives, including long-term leasing of state lands in order to reduce prices of residential units and remove the land price burden. To promote their objectives in the field of production and supply of housing, the Government prepared the Act of Organising and Supporting the Production and Supply of Housing, and submitted it as a top-priority bill on 22 May 2007 to the Parliament, with the Parliament approving it on 14 May 2008 followed by the approval of the Council of Guardians (Housing foundation of Islamic Revolution 2012).

The banking sector, particularly the Maskan (Housing) Bank, has so far given up to 102 trillion Rials ($10.2 billion) of loans to applicants of the Mehr Housing Project. Under this scheme real estate developers are offered free lands in return for building cheap residential units for first-time buyers on 99-year lease contracts. The Government then commissioned agent banks to offer loans to the real estate developers to prepare lands and begin construction projects in an attempt to boost production and balance supply against demand.

In 2010 and 2011 about 1.5 million residential units were built by the Mehr Housing Project, but main questions are whether these housing units solve the problem of housing in Iran and, whether people prefer to live there. One of the main obstacles for many people to buy residential units in the main cities of Iran is the impact of land prices on housing price. The Mehr Housing Project's incentive to build multi-storey residential buildings in suburban areas of the main cities was meant to address this problem. Most of these housing projects were actually built near New towns, thus sharing their problems (lack of services, job opportunities, and identity crisis, also social conflicts, and so on).

The main priority of this programme is providing residential units with creation of job opportunities often being ignored, leaving the question which one should be given priority. We will answer this question using Maslow's hierarchy of needs. Maslow's classification of needs (1970) consists of five levels of cognitive needs, including physiological needs, safety needs, belongingness and love needs, esteem needs, and the need for self-actualisation (Asad Poor and Jusan 2012).

For the most part, physiological needs are obvious as they are the prerequisites of human survival. If these requirements are not met, the human body simply cannot continue to function (Maslow 1954). According to the Maslow's hierarchy of needs, food and water are the most important needs, followed by shelter. If we consider nutrition as the most important need, we have to admit that having a job is the means for most people to have sufficient nutrition in the long term. Also, we have to consider that most immigrants from villages and small towns coming to larger cities had homes in their birthplaces, and it has been the lack of job opportunities that forced them to abandon their own homes and try to land a job in larger cities. Providing job opportunities, therefore, is arguably more crucial than providing residential units.

15.7 Conclusion and Suggestions

Due to the fact that 9,044,823 people (born during 1981–1986 period) have entered or are about to enter housing market in Iran, and also that about ten million people born between 1986 and 1996 will enter the market in the next few years, the housing market in Iran is facing high demands. It seems that the Government has built enough residential units to cope with this demand. The Mehr Housing Project has built 1,500,000 units between 2010 and 2012 alone, and it seems that the continuation of this programme can provide sufficient residential units in the forthcoming years, but the main problem with these units is that they are left vacant and people tend not to live in them. The government, thus, has to take steps to reverse the trend. Below are some suggestions how to encourage people to live in these units:

- Creating job opportunities (especially in manufacturing and agriculture sectors) in new towns and suburban areas where Mehr Housing Project residential units have been built.
- Thorough studies on properly locating new residential developments (ideally close to areas with job opportunities).
- Providing public transport for fast and affordable access to work places and main cities.
- Paying attention to social cohesion issues by encouraging people of the same culture, job, class and religion to live together.
- Providing public services such as hospitals, schools, local security, libraries, artistic cultural centres, and the like.

References

Aronovici C, McCalmont E (2010) Catching up with housing. Newark, New Jersey: Beneficial Management Corporation, United States

Asad Poor ZJA, Jusan MM (2012) Exploring housing attributes selection based on Maslow's hierarchy of needs. Procedia-Social and Behavioral Sciences Vol. 42

Hopkins LD, Zapata MA (2007) Engaging the future: tools for effective planning practices. In: Hopkins LD, Zapata MA (eds) Engaging the future: forecasts, scenarios, plans, and projects. Lincoln Institute of Land Policy, Cambridge, MA, pp 1–17

Housing foundation of Iran website [online]. Available from http://www.bonyadmaskan.ir/En/SitePages/MaskanMehr.aspx. Accessed 13 Aug 2012

International Covenant on Economic, Social and Cultural Rights website [online]. Available from http://www2.ohchr.org/english/law/cescr.htm. Accessed 5 Aug 2012

Iranian Statistics Centre, n.d. [online]. Available from www.amar.org.ir. Accessed June 2012

Islam Online. Available from http://islamonline.com/news/articles/17/140000-Mehr-Housing-Units-Ready-Soon.html. Accessed 15 Aug 2012

Maslow A (1954) Motivation and personality. Harper and Row, New York Chap. 4

Masnick GS (2002) The new demographics of housing. Housing Policy Debate 13(2):275–322

Ministry of Housing and Urban Development (1990) New Towns. MHUD Press, Tehran

Myers D, Ryu SH (2007) Aging baby boomers and the generational housing bubble: foresight and mitigation of an epic transition. J Am Plan Assoc 74(1):17–33

National American daily newspaper published by the Gannett Company [online]. Available from: http://www.usatoday.com/news/world/story/2012–07–29/iran–baby–boom/56576830/1. Accessed 10 Aug 2012

Wikia [online]. Available from http://economics.wikia.com/wiki/Effective_Demand. Accessed 10 Aug 2012

Ziari K (2006) The planning and functioning of new towns in Iran. Cities 23(6)

Ziari K, Gharakhlou M (2008) A study of Iranian New Towns during pre and post revolution

Chapter 16
The Evaluation of State Involvement in Large-scale Property-Led Regeneration Projects in Iran

Alireza Vaziri Zadeh

Abstract During the last two decades a number of state-led, large-scale property-led regenerations have been implemented in Iran. This chapter evaluates the reasons for the failure of these projects by critically discussing two case studies: one in Tehran (Navvab) and the other in Mashhad (the city centre). Mega-scale regeneration projects are usually associated with concerns about the social polarisation and spatial fragmentation in cities. These concerns can be observed more or less in many European redevelopment initiatives. Yet, reviewing mega-scale regeneration in Iranian cases reveals the existence of the same concerns, and more. Using the theory about state-failure, the chapter argues that the additional issues Iranian cities are involved with compared to European cases, is because of the dominance of the state-led approach and the lack of institutional capacity in regeneration programmes, leading to their failure.

Keywords Large-scale regeneration · State failure · Social polarisation · Institutional capacity · Navvab regeneration project · Mashhad city centre

Abbreviations

CBI	Confederation of British Industries
MHUD	Ministry of Housing and Urban Development
ROT	Regeneration Organization of Tehran
URSPIC	Urban Redevelopment and Social Polarisation in the City

A. Vaziri Zadeh (✉)
Faculty of Engineering - Department of Architecture, KU Leuven, Leuven, Belgium
e-mail: alireza.vazirizadeh@asro.kuleuven.be

F.F. Arefian and S.H.I. Moeini (eds.), *Urban Change in Iran*,
The Urban Book Series, DOI 10.1007/978-3-319-26115-7_16

16.1 Introduction

Generally speaking, the term regeneration is ascribed with any kind of redevelopment taking place in cities. Since the 1990s the governmental regeneration projects gained a momentum in many Iranian cities. However, what is called 'the new regeneration movement' today in the UK and the EU, initiated after the 1980s as a result of dramatic neoliberal socio-economic changes in industry-based cities. The approach toward regeneration has been revised during the last thirty years, but the ultimate aim of all projects is to maintain the role and status of the city nationally and globally. In this context the redevelopment of derelict parts has been applied to facilitate new activities and opportunities for cities (Jones and Evans 2008).

Regeneration initiatives, in most European cities, have considerably changed the landscape of cities due to the fact that they have been accompanied by the proliferation of consumption and knowledge-based activities, the growth of worldwide relationships, and changes in the landscape of exclusion/inequalities/deprivation, the increase in socio-cultural diversity, and the emergence of new types of urban spaces in cities (Tallon 2009).

Property-led regeneration has been a prevailing approach in regeneration since the 1980s. This trend is based on the neoliberal consensus that competitive market-based economy could find the most efficient solution to deal with urban problems (Ambrose 1994). The trickle-down effects of the developments are supposed to bring further boost to the micro-economy of the area and the macro-economy of city. The property-led regeneration is known as 'entrepreneurialism' characterised by public-private participation, liberalisation, privatisation, deregulation, and—typically—centralisation (Healey et al. 1992).

Property-led approach is also known by megaprojects: the large-scale, highly speculative, mixed-used flagship projects applied as the symbol of regeneration and a powerful tool to improve confidence for further investment and to define a new opportunity in urban market.

16.2 Megaprojects

The so-called Megaprojects (Fainstein 2008) or Flagship Projects (Bianchini et al. 1992) or Urban Development Projects (Moulaert 2005) are known as significant, high-profile, property developments playing influential roles in urban regeneration. Megaprojects are presented as opportunities to change the level and function of the city in the urban regional hierarchy, to create jobs, and to strengthen the position of the city in the division of labour. They indeed emerged as the main strategy to improve the tax basis of the city by revaluing prime lands to stimulate the economy and entrepreneurial innovation of cities (Healey et al. 1992).

The term 'flagship' is commonly applied to pioneering, large-scale urban renewal projects (Bianchini et al. 1992). The approach toward flagship megaprojects was adopted in the UK, Europe and United States (mostly after the 1980s) as the result of the Government's predilection for physical regeneration and increasingly competitive international context for promoting opportunities in localities. It was assumed (CBI 1988) that every city should have one or more megaprojects to break the circle of city decline.

Megaprojects have signalled intentions for regeneration of derelict areas and act as a magnet for further developments of indigenous attractions. Typically they focus on areas with most development potentials, like heritage sites or waterfront locations. They are going to fully exploit the assets of the cities that are not known beyond their boundaries.

As Swyngedouw et al. (2002) argues, the main objective of megaprojects (such as offices, luxury residential areas, museums, waterfronts, exhibition halls and parks, business centres, and international landmark events) in European cities is maintaining the competitive economic position of the metropolitan in ever-changing global conditions.

Improving the competitive position of the city means creating a new image for urban space primarily in the eyes of investors, developers, businessmen and even money-packed tourists (Fainstein 2008). Thus, megaprojects are applied as tools for 'place-marketing' to attract capital investment.

There are many examples of mega-scale projects in European countries: Adlershof in Berlin, consisting a high volume of offices and housing to respond to the demands of international investors after the unification of Germany (Hāußermann and Simons 2005), Leopold Quarter in Brussels that responded to the additional high demand for office spaces from European Union institutions (Baeten 2005), the Olympic Village in Athena (Delladestsimia 2005), the Thames Gateway and the South Bank in London, and Bijlmermeer in Amsterdam (Fainstein 2008).

16.3 Concerns Regarding Megaprojects

Considering the fact that megaprojects are more than just property investments, they have to be assessed with wider ranging criteria. Moulaert et al. (2005) used the result of URSPIC[1] (Urban Redevelopment and Social Polarisation in the City) regarding megaprojects to argue that megaprojects raise controversy in the following ways:

- The democratic mechanism of local participation usually fails to be inclusive as participation in megaprojects is often limited to those from selected influential groups, and less powerful groups are mainly excluded (also in Tallon 2009).
- Megaprojects' planning vision is often too short-sighted and thus the impacts of the project on the city and its area are very uncertain.
- An excessive focus on design, detail and aesthetics produces a landscape dominated by bulky buildings, in contrast with the context of the city. They

[1]URSPIC is a research about thirteen large-scale projects in the EU.

fabricate fake heritage themes contributing to aesthetic of a standard city (also in Fainstein 2008).

- Given the working of the real-estate market, most of them resulted in deepening social polarisation.
- They are targeted for particular users, too expensive, awkward and unwelcome for local residents to enjoy as consumers of new attractions.
- Megaprojects incorporate a restructuring the local/regional governance. The new structure redistributes power and responsibility to exclusive partnership agencies (mentioned as 'privatisation of urban governance').
- Almost all megaprojects are state-led and often state-financed. The state has to encourage some profitable schemes and apply a variety of methods like low-interest loans, regulatory reliefs, and tax reductions to entice private developers to invest in property and construction. Yet, the state is inevitably the most important player in terms of risk taking.
- Megaprojects are highly risky and their economic achievement and financial viability depends on the production of urban extra rent; therefore a future increase in urban rent is inevitable.

Despite all issues associated with megaprojects, they can break the spiral of city decline by improving internal and external images of cities, adopting the new international/national division of labour, and pursuing consumption revenues in tourism, retail and other service industries.

16.4 Mega-Scale Regeneration Projects in Iran

As Healey and Barrett (1990) argue, in order to understand the particular form of a development, in a particular place and particular time, one needs to find out the set of strategies, interests and actions of organisations and individuals in the context of socio-economical process. Thus, to study mega-regeneration projects in Iran. It is necessary to identify the agents involved, their role, strategies, and their interests within the process of (re)development.

During the 1990s and 2000s, mega-scale regeneration projects implemented by governmental and quasi-governmental agencies are divided into two main categories. The first includes entrepreneurial practices undertaken by the municipalities of major cities to generate income when they become financially (not politically) independent. The central government initiated a policy in the late 1980s as a part of economic reforms to let municipalities of some large cities like Tehran be economically self-supporting, although the mayor was still appointed by the central government. Investing in real estate market through implementing regeneration projects, particularly when they could benefit from monopolisation, could provide them with considerable profits. The Navvab project in Tehran and the Kaboud Mosque commercial complex in Tabriz are the most important projects of this kind during the 1990s (Izadi 2008).

The second category includes the large-scale property-led regeneration around religious centres and important shrines to expand their territories. These projects were mainly imposed and implemented by The Ministry of Housing and Urban Development (MHUD) and strongly supported by the religious establishment.

In general, MHUD and its affiliated organisations are main agents responsible for urban development in Iran. The redevelopments of the areas in the vicinity of the holly shrine, tombs and mosques have been included in the MHUD regeneration programme after 1990. All those projects were planned and implemented through partnership schemes between the government and religious institutions. The economic and political support of religious institutions was of essential importance for the achievements of these projects.

The powers of religious institutions have enormously increased after the Islamic revolution in 1979. They were fundamentally re-organised to be bureaucratically and financially autonomous. These organisations take advantage of exemptions from taxes and many legal restrictions, easy access to various fiscal and monetary facilities and income generated from religious charities (Izadi 2008).

In what follows the regeneration of Navvab district in Tehran (an example of the first category) and the regeneration of historic urban core of Mashhad (an example of the second category) are described in detail.

16.4.1 Tehran; Navvab Regeneration Project

According to Tehran's first master plan (1968) Tehran was divided to isolated regions connected to each other via a network of highways. One long highway was planned to connect the northern part of the city to the new airport in southern outskirts of the city. The highway had to cut the historic core of the city on its way. The northern part of the highway was implemented during the 1970s, yet the extension of the road stopped when it reached to the historic core.

The beginning of 1990s coincided with the flourishing of development projects in Tehran highly supported by the central government. Changing the image of the capital city was one of the state's objectives as Tehran had always been the pioneering role-model for the development of the country in the previous two centuries.

Plans to build the extension of the highway gained momentum in the early 1990s under a large-scale regeneration scheme for the old core of Tehran. In 1992 the master plan of Tehran was revised and the extension of the highway was reconfirmed. This highway was to form the western part of a high-speed ring road surrounding the historic core of Tehran (Bahrainy and Aminzadeh 2006).

The aim of the urban regeneration was mainly changing the blighted image of the old parts of the city with a new and modern image. It was claimed that the trickle-down effects of the development (according to which property development brings in further businesses and jobs to the area and further boost to the economy) would improve the overall quality of life in the derelict core of Tehran (Tehran Municipality 1992). Although the notion seems naive and highly ambitious now, politicians and many technocrats supported it. They considered some high-profile redevelopments during the 1980s (like the redevelopment of Canary Wharf) as good examples and believed that the same approach had to be adopted in Iran (Iran Urban Studies 2010; Tehran Municipality 1992, 1996).

By introducing profitable land-uses (residential, commercial space and offices) and considerably increasing densities of new buildings, the Municipality expected to generate an enormous income from the added value (Iran Urban Studies 2010). It decided to earn all profits of the development by monopolising the investment of the entire project.

The regeneration area amounted to 80 ha with a population of 260,000 people living in 20 neighbourhoods. Although the households in the area were rather deprived, they were socially coherent. Yet, in the absence (or weakness) of a civic society, it was easy to ignore community participation and to impose the regeneration scheme on the city (Bahrainy and Aminzadeh 2007).

The profile of the scheme comprised a highway in the middle surrounded by high-rise dense residential blocks to its sides. The length of the project was 5530 m. More than 8500 new residential flats (with an area of 750,000 m^2) were planned to be built in the area a few of which given to previous inhabitants as compensation (Tehran Municipality 1992, 1996).

In late 1992 the planning of the project initiated. A huge number of properties should be acquired through compulsory purchase. In many previous cases compulsory purchase has been awkward, time-consuming and mostly unsuccessful. Yet in this project the municipality supported by autocratic power and sufficient finance managed to acquire lands in a relatively short time and initiated the implementation in 1994 (Tehran Municipality 1997).

The implementation of the project required a substantial budget. There were three sources to support costs: the general Municipality budget, public bonds issued by the Municipality with a (relatively) high rate of interest (20 %), and the selling by the municipality to the private investors of units in the time of construction at a discounted price. Applying the two latter policies, the Municipality could cover near 75 % of the costs. However, due to fluctuations of the real-estate market and highly optimistic economic feasibility studies, the project encountered long delays in providing sufficient revenues.

After a long delay the implementation of the highway and buildings of the first, second and third phases of the scheme were accomplished by the early 2000s. The regeneration schemes for the fourth and fifth phases were cancelled after governmental changes in the management of the Municipality in the mid-2000s.

Bahrainy and Aminzadeh (2007) argue that there are many issues emerging by the implementation of the project; e.g. the identity of the place was significantly lost due to the stark contrast between the new premises and the old context; the vitality declined due to the loss of public spaces and the security of the area decreased due to the emergence of many indefensible spaces (poor design).

Also, according to a research by Etemad et al. (2012), it is revealed that less than 20 % of previous inhabitants returned to new flats. It means that the regeneration resulted in a massive displacement in the area. Rafiei et al. (2009) state that in average the new households live in new flats for between 4 and 5.5 years while it used to be 18.5 years. They (about 72 %) prefer to move to other places if they can.

16.4.2 Mashhad: The Regeneration of the Historic City Core

The city of Mashhad has developed around the holly tomb of *Imam Reza* during its history. The development of the Shrine in Mashhad has always been used as a 'monumental propaganda' to display the 'prestige' and 'grandeur' of the State (Cavalcanti 1997) (Fig. 16.1).

During the 1920–1930s, a ring road surrounded the shrine to define the sacred place. In the 1970s, the state imposed another extreme intervention on the city core and around the shrine area. In the new development the circular road was offset to a radius of 320 m, leaving a large empty area surrounded by a round wall. The shrine was completely detached from its vicinity (Borbor 1974).

Fig. 16.1 The holy shrine in the Mashhad city centre (A. Bonakdar, 2014. Reproduced by permission of Ahmad Bonakdar)

The redevelopment of the historic core of the city (particularly around the Shrine) intensified after the revolution. The comprehensive regeneration scheme for 337 ha of urban area in the vicinity of the Shrine was planned by MHUD in 1993–1995. The scheme comprised of many large-scale regeneration projects in the area.

The redevelopment scheme aimed to change the old spatial organisation of the area by superimposing new streets. The new straight streets would divide the formerly coherent organic fabric into isolated blocks. Also, the 'obsolete urban fabric' ought to be replaced with new modern (relatively more dense) residential and commercial buildings (Tash Consultant Engineers 1995).

A new quasi-governmental organisation, called as '*Mashhad Housing Development and Construction Company*', was established in 1994 to finance and implement the redevelopment projects. The shareholder of the agency included the MHUD, the Municipality of Mashhad, and the religious administration of the Holy Shrine. Also, to provide the finance for redevelopment projects, mortgage bonds with 17 % annual profit rate were issued.

The land for the projects should be acquired through compulsory purchases. The increase in land value (due to improvement of real-estate market) was considerably higher than the total value the agent paid to the former landowners and the bank interest rate. The price per square metre of the land increased from 250,000 Rls (1992) to 1,000,000 Rls (1999), 1,500,000 Rls (2004) and 4,000,000 Rls (2008). This highly profitable purchase persuaded the agent for acquiring more land; even outside redevelopment area. To enforce people resisting against compulsory purchases and eviction, the agent left many premises abandoned resulting in a decline in the security and quality of the area. This policy forced remaining people to displace and made the land acquisition even easier (ROT 2011).

During 1999–2000 the scheme was revised through the collaboration of the three main shareholders. The new scheme put more emphasis on pedestrians, particularly in the vicinity of the Shrine, while the structure of the former plan did not change.

It should be mentioned that the scheme was divided into three concentric ring zones. The first, central zone is the shrine area. The religious administration (Astan-e-Qods) has been the only agent responsible for the shrine (first zone). The authority of this agent rests on its autonomous economic base and religious and political influence on the central government. It is the owner of a myriad of properties and vast lands all around Iran with an annual budget of $2 billion (Saeidi 2004). By making new buildings, courtyards, porches, roads and streets, the agent has quadrupled the area of the Shrine during the last two decades.

The large-scale regeneration projects of the historic core of Mashhad have been associated with highly centralised structures of governmental power. Although there was a rhetoric of participation of the private finance, in reality governmental authorities revealed little enthusiasm in participating private partnerships.

Although the central government and its affiliated agents have managed the entire redevelopment scheme, the fragmentation, diversity and plurality of agents has been the source of many issues. The absence of constructive co-operation and existence of some political tension between these governmental agents resulted in long-term delay in the implementation of regeneration.

16.5 Evaluation of Key Factors of Regeneration

To evaluate regeneration programmes, some key factors influencing all such projects should be critically analysed. Moulaert et al. (2005) illustrate some common factors characterising mega-scale urban regeneration projects in Europe. They critically analyse thirteen cases to find out how mega regeneration projects have led to economic restructuring and social polarisation in cities. In what follows the outcomes of their research will be discussed, and then applied to the two Iranian cases studied here.

16.5.1 Projects, Not Planning

The mega regeneration projects In European cities are mainly applied as high-capacity strategic instruments remarkably flexible to stimulate economic growth; and they are considered as alternatives for comprehensive long-term planning (the Fordist age policy instrument). These kinds of projects have always been considered as exceptions, which can be implemented through particular privileges, like special governing system. Consequently, they have weakly followed the rules and policies of cities master plans.

A review of the Iranian cases reveals the same situation. In Mashhad the core of the city was regarded by MHUD as an exceptional case to be planned through specific redevelopment projects. In the latest comprehensive development plan of the city, the area represented as a black spot which means it is a specific zone that is not included in the comprehensive master plan (Farnahad 2005). In Tehran, the decisions regarding the Navvab regeneration project was made by an exclusive committee in 1993, regardless of the policies of Tehran master plan.

16.5.2 Development Agents

The diversity and complexity of the agents in different European large-scale redevelopments is very significant, from grassroots to public organisations; e.g. In Berlin (Adleshof) the institutional organisation is based on the partnership between the public sector and semi-private developers (Hāußermann and Simons 2005). In most cases there is a sort of participation between public and private organisations that is accompanied by a limitation of the state intervention to accommodate the requirements of private institutions.

In Iran the agents of large-scale redevelopments are based on a consortium of the central government organisations (mainly MHUD), municipalities and in some cases religious institutions. In many cases, the key decisions are usually taken in governmental meetings. The roles of private organisations are limited by the wills and policies of the governmental agencies. In fact the government decides when and how the private sector can participate in projects.

16.5.3 Funding; Public or Private

In almost all recent European megaprojects, a specific neoliberal approach toward development policies can be observed. Investors try to exploit the area by producing as much high rent-yielding spaces, e.g. luxury residential housing and offices, as possible. This approach is to reorient the state monopolisation in market regulation toward providing infra-structure. Consequently, in many cases the state has withdrawn to a greater or lesser extent. However, in contrast with the rhetoric of the market-led essence of entrepreneurs, almost all projects are state-led and it is the state that takes the main risk.

In Iran, however, the role of the state is much more dominant. The state is not the only entity making almost all decisions and taking the risk, but also the key financier and developer. The main reasons why the private sector does not participate in financing in large-scale projects are as follows:

- Due to a highly monopolised system of decision making and the right of confiscation by the government, private investors (particularly those not so close to the government) do not put their trust in investment. Thus, the government should entice the private capital by only publishing public bonds;
- Taking the advantage of oil revenues, especially in periods of sudden increases in prices, leaves the central government confident that they do not need private capital to support investment in costly, large-scale projects.

16.5.4 Participation and Accountability

For the sake of efficiency and flexibility in institutions and agencies managing urban regeneration, in most European cities new forms of governing connecting the state and the civic society is welcome. In such arrangements, participation is the norm as it permits communication between players; and consequently in this form of urban governance, accountability, representation and inclusion should be respected. However, in reality in some cases the channels of accountability have been non-transparent and unregulated; e.g. in the Olympic Village in Athens (Delladestsimia 2005).

Reviewing the two cases in Iran reveals that projects are based on a top-down autocratic approach supported by a group of elites with no evidence of accountability. The lack of accountability originated from the general political governmental system in Iran in which the state assumes itself the right to impose its plan as good will on society (Mazumdar 2000). Corruption can normally be expected in the absence of accountability. Generally, when there is some kind of corruption in a development programme, it becomes vulnerable to failures. Also, there are many issues regarding community participation in regeneration programmes. Communities living in such areas have little chance of engaging in regeneration projects while the organisations who are the agents of regeneration never make any effort to create the capacity required for community involvements.

16.5.5 Territorial Fragmentation

Most large-scale regeneration projects in European cities have contributed in the fragmentation of cities. It means that the regenerated areas have not severed ties functionally and physically from adjoining deprived neighbourhoods. Some regeneration programmes have resulted in emerging gated communities as middle-class areas were completely detached from their surroundings (Tallon 2009; Porter and Shaw 2009). For example the Olympic Village in Athens and Leopold Quarter in Brussels resulted in this kind of detached areas.

The same can be observed in Iranian regeneration programmes. Both studied cases are detached from their surroundings. In Tehran, the highway makes a sharp edge between east and west sides of the regenerated area cutting many connection routes. In Mashhad, due to the fact that the regeneration area was considered as the new territory of the sacred place, the detachment of the new area from the surroundings was explicitly one aim of the programme. To achieve this, a ring road has created a boundary around the regeneration area.

16.5.6 Social Returns

The social return of regeneration projects is one of the most important concerns for recent European programmes. Many regeneration programmes have included social initiatives, focusing on issues such as crime, unemployment, poor health and education (Jones and Evans 2008). It means that these programmes should comprise initiatives to the benefit of the deprived households directly. For example the Olympic Village in Athena included some initiatives to benefit grassroots like training people during construction and the adaptation of the village as social housing after the games (Delladestsimia 2005).

Yet, large-scale regenerations in Iran have rarely considered the social return of the projects despite the fact that the main agent for regeneration is the government. As can be seen in the Tehran and Mashhad cases the programmes do not include any initiatives to benefit deprived households in the surroundings. Regeneration agencies are not committed to provide any social houses or producing jobs for the public benefit.

As mentioned, megaprojects have been accompanied by issues related to the distribution of development benefits throughout the society. Fainstein (2008) believes that the extent to which the gains of any development can be spread into cities depends on how much the government directly demonstrates commitment to the public good. In some cities where there is a social-democratic ethos and a welfare government (for example in Amsterdam), the social returns of these projects are greater and the social-economic issues coming out of large-scale redevelopment are smaller.

Yet, a review of the Tehran and Mashhad cases reveals the fact that the same issues associated with large-scale projects in European countries can be observed in Iran However, the issues appear much more intensive as the projects resulted in social polarisation and exclusion much more than their European counterparts. Compared with European cities, the projects in Tehran and Mashhad are at considerably lower levels of accountability and participation. However, they depended on public funds more than European initiatives, they have produced much less social returns and resulted in more fragmentation in the context of their cities.

Given the fact that a particular form of a development, in a particular place and particular time is the outcome of interactions between development agents in terms of their strategies, interests and actions (Healey and Barrett 1990; Madanipour 1996), it can be claimed that the main reason for all above-mentioned concerns is the relationship between the main agents (actors) of regeneration projects. Obviously, in all large-scale redevelopment projects in Iran, the role of the state is highly prevalent. While the state is the key player controlling the process of redevelopment, it has accepted little commitment for the public good. The state plays as an entrepreneur who seeks benefits and rents from urban development projects through monopolisation of the market for its own organisations. In other words, the state fails to fulfil the main responsibilities it should do in development programmes; e.g. the correction of real-estate market through dedication of social housing for low and middle classes, the distribution of welfare by providing jobs and training and the stabilisation of property rights by decreasing the risk of eviction and expropriation.

To explain why the state in Iran could not fulfil the responsibilities it is supposed to, I refer to the theory of state failure in developing countries. This theory explains the reasons and conditions in developing countries that result in the emergence of the issues associated with failure of state-led development programmes.

16.5.7 The Theory of State Failure

According to Khan's (2004) argument about state failure, particularly in developing countries, there are two points of views that explain the reasons why the state in developing countries like Iran could not fulfil the responsibilities it must.

The first called 'service delivery' argues that the role of the state is to provide 'order and law', 'correction of market failure', 'essential public goods' 'welfare distribution', and 'stability of property right'. The failure of the state to do these, results in economic under-performance and disruption of development. In this view, there are three 'core components' whose coexistence result in development programmes failure. These components are: the instability of property rights and distortion of market due to state-created rents, rent-seeking and corruption, and the absence of democracy.

The second view called 'social transformation' considers the role of the state in the context of the transition into industrial-capitalism from pre-capitalisation and pre-industrialisation. During the transformation, the state (in developing countries) should highly intervene in property rights and rent management systems to speed up acquisition of modernisation and capitalist transition.

The failure of the state according to this view is driven by the lack of institutional capacities, and more importantly, the incompatibility of institutional capacities with pre-existing distributions of power. Khan (2004: 2) argues that 'the more critical area of state failure has been the absence of adequate institutional and political capacity in developing country states to assist and accelerate a dynamic transformation'.

In developing societies, the state inevitably utilises its power and violence going through economic and social changes. In this situation, the state is decidedly vulnerable to be used as a tool in the hands of contending classes and political entrepreneurs who try to capture resources and divert the transformation into specific directions. In this view, state failure can be prevented if 'democracy, civic society participation and other desirable political institutions could be deepened' (Khan 2004: 35).

16.6 Conclusion

The role of the state in large-scale redevelopments is of vital importance: the state is the only entity capable of legitimately (re)distribute income and wealth, represent social cohesion, resolve conflicts, and collect tax, and in all cases applying force if necessary. Thus it is the state that has to protect the stability of property rights during any urban redevelopment programme, particularly in developing countries. It should ensure the society of low rents, low corruption and undistorted market. However this would not be possible unless there is civic society participation and democratic accountability to keep the state in constant control (Khan 2004).

Given that the state is the dominant, even the sole, actor in regeneration initiatives in Iran, reviewing the evidences of these cases reveals the fact that the considerable amount of rent producing (by the state) through implementation of large-scale redevelopments has been accompanied by instability of property rights (compulsory purchases or expropriation of the properties) and distortion in real-estate market (displacement of indigenous inhabitants owing to increasing of housing rents). The state-created rents were incentives for rent-seeking by groups of special privilege and power (like quasi-governmental organisations or religious foundations). In the absence of democracy they could monopolise political power and benefit from considerable rents of redevelopment projects. When this was accompanied by corruption and politically appointed supervisory administrations, it resulted in an intensification of the issues associated with large-scale developments.

From another point of view, the state in Iran has pushed the society and cities toward modernisation and industrialisation since the 1920s. It can be claimed that large-scale redevelopment projects have been flagship initiatives to achieve the urban development and transition to modernisation rapidly. Yet, in a situation of a weak democracy, inefficiency of civic society participation and lack of many desirable political institutions, regeneration projects are used as a tool for rent-seeking.

Overall, the reasons why mega-scale state-led regeneration programmes during the last two decades could not achieve their aims and now raising many concerns can be summarised as:

- The state monopoly in making decision and management;
- The lack of accountability and democratic processes;
- The lack of appropriate institutions;
- The lack of involvement of local communities;
- The lack of efficient participation with the private sector.

References

Ambrose P (1994) Urban process and power. Routledge, London

Baeten G (2005) Old elites in a new city. In: Moulaert et al (eds) The globalised city. Oxford University Press, Oxford

Bahrainy H, Aminzadeh B (2006) Autocratic urban design: the case of the Navvab Regeneration Project in central Tehran. Int J Environ Res 1(2):114–127

Bahrainy H, Aminzadeh B (2007) Evaluation of Navvab Regeneration Project in Central Tehran, Iran. Int Dev Plann Rev 29(2):241–270

Bianchini F, Dawson J, Evans R (1992) Flagship projects in urban regeneration. In: Healey P et al (eds) Rebuilding the city: property-led urban regeneration. E & FN Spon, London

Borbor D (1974) Iran. In: Whittick A (ed) Encyclopaedia of urban planning. McGraw-Hill Book Company, New York

Cavalcanti M (1997) Urban reconstruction and autocratic regimes: Ceausescu. Plann Perspect 12:71–109

Confederation of British Industries (CBI) (1988) Initiatives beyond charity. CBI, London

Delladestsimia PM (2005) The olympic village. In: Moulaert et al (eds) The globalised city. Oxford University Press, Oxford

Etemad G et al (2012) Arzyabi-e-Tarh-e-Navvab va Payamadhay Aan (Evaluation of Navvab Project and its consequences). Maani Publisher, Tehran

Fainstein S (2008) Mega-projects in New York, London and Amsterdam. Int J Urban Reg Res 32:768–785

Farnahad (2005) Tarh-e-Tosee` va Omran Nahiyeh-e-Mashhad (Development plan of Mashhad region). Mashhad

Hāußermann H, Simons K (2005) Facing fiscal crises. In: Moulaert et al (eds) The globalised city. Oxford University Press, Oxford

Healey P, Barrett SM (1990) Structure and agency in land and property development processes: some ideas for research. Urban Stud 27:89–104

Healey P et al (eds) (1992) Rebuilding the city: property-led urban regeneration. E & FN Spon, London

Iran Urban Studies (2010) Navvab, Eshtebahi ke nabayad tekrar shaved (Online). Available from www.iranurbanstudies.com. Accessed 5 June 2012

Izadi M (2008) A study on city centre regeneration: a comparative analysis of two different approaches to the revitalisation of historic city centre in Iran. Ph.D. thesis, Newcastle University

Jones P, Evans J (2008) Urban regeneration in UK. Sage, London

Khan M (2004) State failure in developing countries and institutional reform strategies. In: Tungodden B, Stern N, Kolstad I (eds) (2005) Toward pro-poor policies. Aid, institutions, and globalization. Annual World Bank conference on development economics, Europe. 2003. Oxford University Press and World Bank, pp 165–195

Madanipour A (1996) Design of urban space: an inquiry into a socio-spatial process. Wiley, Chichester

Mazumdar S (2000) Autocratic control and urban design: the case of Tehran, Iran. J Urban Design 5:317–338

Moulaert F, Rodriguez A, Swyngedouw E (eds) (2005) The globalised city; economic restructuring and social polarization in European cities. Oxford University Press, Oxford

Porter L, Shaw K (eds) (2009) Whose urban renaissance?: An international comparison of urban regeneration strategies. Routledge, London, New York

Rafiei et al (2009) Sanjesh-e-Mizan-e-rezayatmandi sokonati-e sakenan Navvab, vol 67, pp 53–68 (Online). Available from www.urbanmanagement.ir. Accessed 5 June 2012

Regeneration Organization of Tehran (ROT) (2011) Tajrobe-e-Mehvar-e-Amir Almomenin: Mashhad (AmirAlmomenin Boulevard Project in Mashhad). Tehran

Saeidi AA (2004) The accountability of para-governmental organizations (Bonyads): the case of Iranian foundations. Iran Stud 37:479–498

Swyngedouw E, Moulaert F, Rodriguez A (2002) Neoliberal urbanization in Europe: large-scale urban development projects and the new urban policy. Antipode 34:542–577

Tallon A (2009) Urban regeneration in the UK. Routledge, London

Tash Consultant Engineers (1995) Tarh-e-Nosazi va Bazsazi-e-Baft-e-Piramoon-e-Haram Motahar (Regeneration of holy shrine). Mashhad, Iran, MHDCC

Tehran Municipality (1992) Report on Navvab reconstruction. Consulting firms (Bavand, Atek, Keyson and Neghsh-e-Jahan). Tehran

Tehran Municipality (1996) Report on Navvab Project. Deputy of Engineering Affairs. Tehran

Tehran Municipality (1997) Report on the evaluation of people's satisfaction in Navvab project. Deputy of Social and Cultural Affairs. Tehran

Chapter 17
Getting Ready for Urban Reconstruction: Organising Housing Reconstruction in Bam

Fatemeh Farnaz Arefian

Abstract Despite optimistic expectations of what a reconstruction programme should deliver, in the delivery of such expectations in practice proved difficult in general and more challenging within urban reconstruction. At present, there is a gap in theory or best practice for organising urban reconstruction at global level. Traditionally, organising complex reconstruction activities as multi-organisational activities within traumatic and complex contexts has been relying on the development of know-hows on a trial and error basis at international level. An example is the development of know-how on organising reconstruction in Iran. As a result of organisational continuity and extensive reconstruction experience Iran was able to apply lessons from each reconstruction to the next. This led to a set of broad reconstruction policies. The Bam reconstruction, after the disastrous earthquake on 26th December 2003, was, however, the first in its kind after such experiences. It was a relatively large city, suffered from heavy human losses in a concentrated zone; it was a historic organic garden city with traditional architectural characteristics. The Bam housing reconstruction programme pursued three objectives: people would be mobilised and participate; buildings would be earthquake resistant; and designs would safeguard the cultural identity and architectural fabric that Bam was famous for. This chapter examines organising post-disaster housing reconstruction programme in the Bam urban context from the perspective of organisation theory. It advances theoretical knowledge on organising urban reconstruction and leads to improvements in organising reconstruction activities in practice. The Bam case presents both strengths and weaknesses that potentially inform other reconstruction cases too.

Keywords Disaster management · Urban reconstruction · Reconstruction programme · Systematic approach · Strategic management · Post-disaster reconstruction · Bam · Iran

F. F. Arefian (✉)
The Bartlett Development Planning Unit, UCL; Silk Cities Platform, London, UK
e-mail: ucfufaa@ucl.ac.uk; f.arefian@silkcities.org; info@civitasphoenix.com

F.F. Arefian and S.H.I. Moeini (eds.), *Urban Change in Iran*,
The Urban Book Series, DOI 10.1007/978-3-319-26115-7_17

Abbreviations

BAUC	Bam Architecture and Urbanism Council
BHRC	Building and Housing Research Centre
BHRP	Bam Housing Reconstruction Programme
BTID	Building Technical Identification Document
CMI	Chartered Management Institute
DRR	Disaster Risk Reduction
HFIR	Housing Foundation of Islamic Revolution
KEO	Kerman Engineering Organisation
MHUD	Ministry of Housing and Urban Development

17.1 Introduction

Iran is of the most disaster prone countries in the world. According to the National Report of Iran, presented in Hyogo in 2005, deadly earthquakes claimed more than 180,000 lives during the last century, and 35 % of Asian earthquakes in the 20th century occurred in Iran (Government of Iran 2005). Contemporary advances in disaster management field characterise post-disaster reconstruction as a facilitative process for delivery of houses instead of mere replacing the lost assets: a process that must integrate people's participation and measure future disaster risk reduction (DRR). However, the reality of organising reconstruction to achieve those expectations has proven difficult. A review of various international examples[1] shows that researchers as well as practitioners widely highlighted practical difficulties for organising and delivering reconstruction programmes in global practice (e.g. Quzai 2010; Davis 2007; Ozerdem and Jacoby 2006; Barakat 2003; Pelling 2003; Alexander 2002; Maskrey 1994). The importance of organisational design for reconstruction programmes is an emerging discourse in disaster fields (Arefian 2015; Davidson 2009; Johnson 2007). Traditionally, organising reconstruction activities as multi-organisational activities and part of the extreme event within traumatic and complex context (Johnson 2007; Comfort and Kapucu 2006) has been relying on developing know-hows on a trial and error basis at international level. An example is the development of know-how on organising reconstruction in Iran.

The country's experience in post-disaster reconstruction was largely enhanced by the lengthy large-scale post-war reconstruction experience for 3891 villages and a number of cities in five provinces alongside 1352 km from 1982 onwards (Meskinazarian 2011). The Housing Foundation of Islamic Revolution (HFIR) is the reconstruction executive body in the country that also deals with rural

[1]For example, reconstruction after Kashmir earthquake, Pakistan 2005, Gujarat earthquake, India 2001, Marmara earthquake, Turkey 1999, Kobe earthquake, Japan 1995, Alto Mayo earthquake, Peru 1990.

development and affordable housing in normal situations. As a result of extensive experience in reconstruction and organisational continuity, Iran could apply lessons learned from each case to the next. This led to organisational learning and establishing broad policies on organising reconstruction, for example, managing the scale of the reconstruction and people's participating, at the time of the Manjil earthquake, in 21 June 1990, and reconstruction.

The 6.5 Richter Bam earthquake, in 26 December 2003, caused around 30,000 fatalities and 20,000 injuries and made 75,000 people homeless. Approximately 85 % of the houses, commercial units, educational facilities and health facilities, as well as public and private offices in Bam and its surrounding villages were completely destroyed or became unusable (UN 2004). Despite having extensive experiences in reconstruction and especially successful Manjil rural reconstruction, the reconstruction of Bam was indeed the first in its kind. Significantly, the earthquake had a shallow epicentre and caused severe destruction, occurring in a relatively large city with historic and cultural traits and causing heavy human losses and casualties in a concentrated zone (Joodi 2010) (Fig. 17.1). Organising housing reconstruction programme in Bam had to build on the previous experiences and policies and fit the situation: a large-scale urban reconstruction within a traditional urban context.

This chapter examines post-disaster housing reconstruction programme in the Bam earthquake from the perspective of organisation theory. Recently, Arefian (2015) suggested that multi-organisational reconstruction programmes require

Fig. 17.1 The post-earthquake Bam city centre (Author 2004)

innovative yet strategic programme delivery system configuration and management. Building on this, the chapter contributes to both theoretical advances and practical improvements for organising reconstruction activities, adopted from a qualitative research.[2]

17.2 Previous Urban Reconstruction Experiences

Iran's experience in both post-disaster and post-war urban reconstruction showed that classic architecture and urban studies are not sufficient for reconstruction as they should incorporate a number of other policies that are not usually part of those plans. Even urbanites are different than villagers, who expect less and are more tolerant of the harsh environment (Zargar 2004). The post-war reconstruction showed that urban reconstruction is more complex than rural reconstruction, which was easier and more manageable. Whilst organising reconstruction is difficult in general, it is acknowledged that urban reconstruction is even more challenging than rural reconstruction. Regulatory frameworks in cities are more complicated and varied in their bureaucracy or efficiency (Shilderman 2010). However, at present there is a gap in theory or best practice for organising urban reconstruction at global level. An example was the reconstruction of Khorramshahr, a modern important port before the war: a flagship project with national importance. Although Khorramshahr was not relocated but saw large-scale physical interventions, without attention to land ownership and inability to acquire required lands for implementing the new urban development plans (interviews 2013; Alizamani 2012; Meskinazarian 2011).

The post-earthquake Manjil reconstruction reconfirmed such intrinsic challenges in urban reconstruction. The earthquake also affected a number of small towns (e.g. Manjil and Roodbar). Urban housing reconstruction in these towns encountered more difficulties in comparison with rural reconstruction. The initial reconstruction programme indicated that all rules and procedures for urban housing were to be followed. Since those small cities did not have enough engineers and architects

[2]Primary and secondary data used for the research included: 59 in-depth interviews with key and knowledgeable people in organisations formulated and implemented the Bam housing reconstruction programme (BHRP) and locals; public project advertisements for the programme procurements in 2004; real-time weekly work progress reports, 2004–2009; the author's direct frequent observations as a participant in implementing the housing reconstruction programme and an urban design project from 2004–2006 (in the capacity of a consultancy director for working with 2,100 families on architectural and technical matters for reconstruction their houses, and an urban design project), followed by observations in 2008, 2010, 2011, and 2013; the real-time reconstruction Guidance for Reconstruction for residential, retail and public units produced by HFIR for locals; official websites of the participant organisations; Bam new post-disaster urban development project documents; publications; selection of the official meeting minutes that were publicly available; the real-time initial typologies for the housing reconstruction by participant consultancy companies in a dedicated exhibition in February 2005, Tehran; and publications.

working with people to follow design and supervision procedures in the early 1990s, The Ministry of Housing and Urban Development (MHUD), called on individual engineers and architects, commissioning them from nearby bigger cities and Tehran. However, in many cases engineers actually did not reside in these cities, and some of them provided drawings without visiting the sites, and there was a minimum level of construction supervision. Reflecting on this experience later, the reconstruction manager[3] suggested that it was probably worth examining the idea of having engineering practices in these cities to assist people with reconstruction (Akhoondi and Bahraini 2000).

17.3 An Innovative Experiment

One of the most difficult tasks in post-disaster reconstruction is organising the necessary processes and procedures, particularly regarding participants' contribution in housing reconstruction. Practical decisions had to be made, in a context of competing interests in emotionally overwhelming post-disaster environment (Davidson 2009). Organising the Bam housing reconstruction programme was based on purposefulness and systematic approaches, characterising it as an innovative experiment and resembling urban development programmes. The housing reconstruction programme was a collection of around 25,000 housing construction projects all following the same standardised process. The systematic approach created a temporary housing development process that aimed to achieve three objectives as the following:

- Safeguarding cultural identity and architectural fabric;
- Constructing earthquake resistant buildings; and
- Mobilising disaster-affected people and their participation.

This experiment shows strengths and weaknesses that can inform the future practice at both national and international levels.

17.4 Strategising the Reconstruction of Bam: Linkages with Urban Development

Natural hazards turning to disasters reveal failures in development activities (e.g. Benson and Twig 2007; Wisner 2004; Davis 2007). Similarly, the Bam earthquake exposed weaknesses and shortfalls. At present the need for developmental approach

[3]The former president of HFIR at the time of the Manjil earthquake, Dr. Akhoondi. He is credited as the architect for establishing current reconstruction policies.

towards recovery and reconstruction is acknowledged in disaster studies. A disaster is a disruption in development path of an area. Post-disaster recovery and reconstruction help to overcome this disruption and the disaster-affected area to return to its development path. There is a spectrum of views about reconstruction from seeing it as a window of opportunity to deliver radical changes, to more pragmatic views that see any situation as opportunities for improvement (See for example, Christopolos et al. 2006; Davis 2007; Johnson 2007; Wisner 2004). The Bam reconstruction approach was defined as developmental and realistic. The goal of the reconstruction as a whole was set as:

> Reconstructing residential and commercial units and other physical assets in cities and villages that were damaged as a result of the Bam earthquake, in line with regional development plans, within the national and local capacities and based on the available capacity and resources in order to return physical life to cities' (Steering Committee for Reconstruction of Bam 2004).

The introduction of the aforementioned objectives within the overall reconstruction goal were linked with the initial consultations on reconstruction, which addressed three aspects of urban and housing development. The above discussions led to introducing three core principles for the Bam reconstruction as a whole.

Firstly, the future of urban characteristics in Bam, given its previous organic traditional garden city characteristics and architectural style, highlighted. Various stakeholders e.g. Iran Cultural Heritage Organisation; MHUD; HFIR; the Bam Town Council, shared the concern about its importance in reconstruction. The Bam case drove attention to historic cities the existing quantitative urban planning system failed to protect. A window of opportunity emerged for change (e.g. interviews, 2013; Beheshti 2004; Esmailie 2004).

Secondly, the severity of destruction proved the inadequacy of existing measures for safe construction. Specialist examinations confirmed that such extensive destruction occurred because the existing seismic codes were not applied during the construction of even newer buildings, although they were compulsory (BHRC 2004; Moghadam and Eskandari 2004). Bam was seen as a small example of a larger threat. Had the earthquake struck another city, the result might have been worse (Sartipi 2006; Hanachi 2004).

Thirdly, previous reconstruction experiences proved that people's active participation in rebuilding their own houses is a win-win scenario for both government and people. This became a main existing reconstruction policy after the Manjil experience, as it could save time and money, deliver houses that people accept and help them to return to normality, and overcome the aftermath trauma. It also showed that instead of delivering turn-key houses the government should facilitate and supervise housing reconstruction by people (Joodi 2010; Akhoondi and Bahraini 2000; Ghafori-Ashtiany 1999).

17.5 Purposeful Approach for Organising Multi-objective Programme

The approach for organising this purposeful and objective-oriented housing reconstruction programme was to organise it around its three objectives. Related policies, practical considerations and required participant organisations identified to achieve each objective, translating objectives into operational levels.

However, achieving objectives requires managing strategically, which relies on strategic thinking (Armstrong 2009). In this regard the Bam case was exercised as a purposeful multi-organisational activity not a full application strategic management. Thus, it showed signs of strategic thinking but it also lacked some features required to call that way of organising the programme 'strategic' (Arefian 2015). The following discussion elaborates an example of this. Strategic management, as the general management of purposeful activities, is about a sense of purpose and looking ahead. It is about organisational configuration, resource allocation and adaptations (for example see: Armstrong 2009; Hanaggan 2002; Ansoff 1990). Keeping the core direction between various levels of strategy and cascading vertically and horizontally is a central element in strategic decision making (Armstrong 2009). Higher level objectives should accurately be translated to lower practical and operational considerations for harmonising of the considerations horizontally at each level aligns all considerations towards the core direction as the main purpose of the activities (e.g. Armstrong 2009; Morgan 2006; Minzberg and Ghoshal 2003; Hannagan 2002; Asnoff 1990).

Given the above, it appears that the existing know-how on organising reconstruction in Iran led to the application of the concept of vertical cascading in order to organise the programme. Using the objectives as reference points helped the reconstruction executive body to cope with the complexities of urban reconstruction, identify required participant organisations, and define roles and responsibilities. The objectives were translated into operational arrangements as the following.

To address people's participation, the housing reconstruction programme was defined as owner-driven. People were responsible for the reconstruction of their houses, and to do so they were financially, technically and administratively supported by the government. A representative from each household was the applicant, the driving force for starting the process, preparation of architectural and technical plans, selecting the contractor and cooperation with engineer supervisors, receiving reconstruction funds instalments and paying the contractor based on the progress of the construction operation (Saemian and Erfanian Daneshvar 2011; HFIR 2004).

To achieve the objective on constructing earthquake resistant buildings, new controlling mechanisms and construction supervision were introduced, and the responsibility was delegated with the Kerman Engineering Organisation (KEO). The KEO is the provincial branch of the Engineering Organisation, the country's regulatory professional body for construction profession. A multi-layer supervisory network was established to control technical plans, as well as construction supervision by the engineers of the KEO. Laboratories for quality control of construction materials (e.g. soil mechanic, cement and brick) and to support inspectors (Havaie and Hosseini, n.d.; Saemian and Erfanian Daneshvar 2011; Steering Committee for Reconstruction of Bam 2004; HFIR 2012).

To achieve the objective of safeguarding urban architectural identity, as well as support HFIR on urban management matters, the Bam Architectural and Urbanism Council (BAUC) was formed from distinguished architects and organised by MHUD. It was formed to ensure that all activities in Bam were design-based. For the city reconstruction as a whole, the BAUC had to identify the historic identity of Bam and develop architectural and urban design guidance. To support beneficiaries HFIR commissioned engineering consultancy companies to establish their local branches in Bam and provide free design and technical services for them. Before the reconstruction began consultancies developed conceptual typologies, based on a variety of locations, directions and sizes; these could be personalised later. Another consultancy company, called the Mother Consultancy oversaw and harmonised the work of these consultancy companies and linked them to the BAUC (Havaie and Hosseini, n. d.; Saemian and Erfanian Daneshvar 2011; HFIR 2012). The methods of engaging consultancies resembled a parallel idea suggested after Manjil reconstruction.

However, despite the merits of vertical cascading, there were some missing elements in the full application of strategic management, which prevented taking the advantage of such an approach during implementation. What was missing was that the three objectives were treated in isolation. In reality, however, they were not isolated. Consequently, mutual agreements and understandings between participant organisations that worked with each other at the same horizontal operational level were not established and priorities were not determined before the programme implementation. These caused implications and unforeseen bottlenecks within the process during implementation, which at one point caused severe delays and frustration, requiring urgent solutions.

One major challenge emerged at the design phase: a conflict between the two controller organisations caused a bottleneck upon the programme implementation. The KEO imposed further restrictions on the existing seismic regulations by converting some optional codes to compulsory. This meant that many architectural design typologies developed earlier under the initial architectural guidance became unacceptable under the new KEO's restrictions. Designs under the initial Mother Consultancy's guidance were not acceptable by the KEO and vice versa. Consequently, severe delays at one point caused frustration and growing backlog. Attempts by the reconstruction executive body to resolve the issue were fruitless. Eventually, the urge to speed up the process—due to people's despair (as they lived in their temporary accommodations) led the Mother Consultancy to compromise. The long list of the initial BAUC's guidance shrunk to a short one for housing reconstruction, namely, to locate the new houses exactly at their previous locations (to protect the garden city characteristics), and to use various textures of brick for facades and to save prefabricated temporary accommodations as secondary supporting spaces (interviews, 2013). The KEO's change of approach largely mono-typed the pocket volume of reconstructed houses. Figure 17.2 presents examples of typical reconstructed houses.

Fig. 17.2 Typical examples of reconstructed houses in Bam and Baravat (Author 2009 and 2013)

The Bam case lacked to identify the overlapping areas between practical considerations for achieving objectives. But in reality they were related to the same production chain for the same product: a house. The Bam case shows how multi-objective reconstruction programmes require more attention to horizontal harmonisation because they require more horizontal operational interconnections between probably more participant organisations, all increasing complexities. In turn growing complexities increase the risk of failure for a system, the likelihood of bottlenecks and unforeseen problems (Bonabeau 2007), as it happened in Bam. Given the already uncertain and complex post-disaster situation all such interconnections must be identified and dealt with accordingly before the programme implementation.

17.6 A Modified Temporary Housing Development System and Process

Initially, it was expected that a large number of households (around 25,000) would want to reconstruct their homes at once. A systemic mechanism was thus needed to avoid chaos (interviews, 2013). In response, a delivery system was formed that combined the practical considerations for achieving reconstruction objectives, reconstruction funding and administrative policies within the overall housing development process. This delivery system, or the 'Model', as called inside the country, was an innovative experiment and a complex system for dealing with urban housing reconstruction at scale. Through standardisation of this the system a new temporary housing development process was created in effect: a modified housing development for all urban houses constructed in the whole urban area. Similar to the overall housing development process throughout the country, the reconstruction process consisted of two separate phases: design and construction (Fig. 17.3).

The approach to the formation of the delivery system was clearly a collaborative attempt to improve the existing housing process tackling weaknesses identified after the earthquake. For example, under normal circumstances no qualitative design codes even existed for promoting traditional architectural identity, and the number of control points at the construction phase increased from five in normal situation to nine. Also, all architectural and engineering documents and construction supervisors' reports were logged to the Building and Technical Identification Document (BTID) which was normally unusual. Interestingly, despite their different opinions on the Model's performance all interviewees for the main research expressed such systematic approach as a valuable feature of the Bam housing reconstruction programme. Formation and implementation of the Model required a gathering of several participant organisations in addition to applicants from policy level and the system formation to very operational level and the programme implementation:

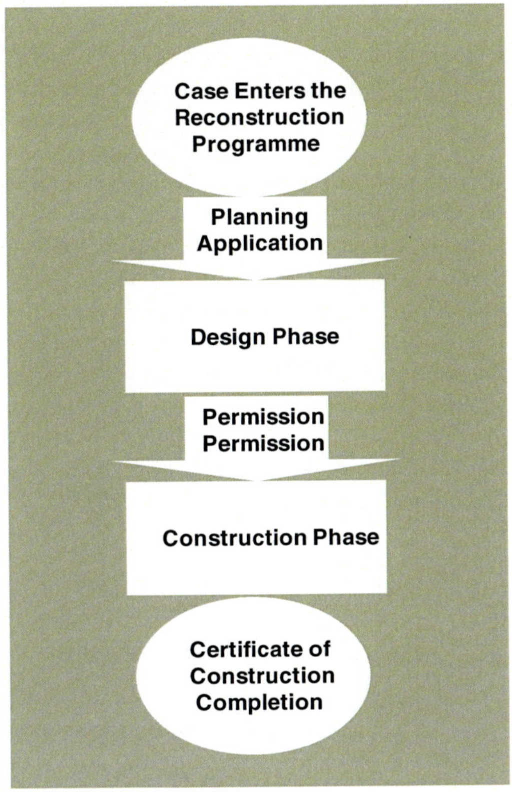

Fig. 17.3 The overall housing reconstruction process in Bam (Author)

HFIR, MHUD, BAUC for the former, followed by the KEO, Setads, Mother Consultancy Company, Municipalities of Bam and Baravat, Local Consultancy Companies, Beneficiaries, Contractors, Banks, BHRC, Laboratories for Quality Control; Construction Material Suppliers for the implementation ad design phase and construction phase.

Working within the overall housing development process simplified the formation of the delivery system for approaching the objectives. It also contributed to longer term improvements. For example, after the official end of reconstruction programme, the KEO retained its role and remained responsible for controlling the structural plans and construction supervision, collaborating with the municipality whereas before the earthquake it was not an influential player in housing development process. Engaging and delegating the responsibility with the KEO contributed to mainstreaming new construction supervision mechanism after the reconstruction period (interviews, 2013).

Working within the overall existing housing development process, however, also means pragmatism and avoiding radical fundamental changes (even improvements) for the purpose of reconstruction (Arefian 2015).

To implement the programme through its delivery system and to manage the scale of urban housing reconstruction which was over 10 times of the 2000 urban cases of Manjil, the Bam urban area was divided to 11 administrative zones with. The responsibility of each zone given to a provincial branch of HFIR as an assistant organisation, called 'Setad'. This approach was adopted from a policy successfully implemented in the post-war reconstruction, to deal with the huge workload, and established in Manjil reconstruction.

Such unit grouping appears to be helpful for dealing with organising large-scale reconstruction. It is replicable, scalable, and potentially facilitates achieving mutual adjustments between participant organisations and individuals during implementation. Each zone was designated to one Setad, one local consultancy and one engineer construction supervision team to serve around 2000 households (applicants/beneficiaries) in that zone and implement the above delivery system (Alizamani 2012). This zoning and operational considerations, in fact, was unit-grouping based on production chain, that is an important organisational configuration (Mintzberg and Ghoshal 2003).

17.7 A Closer Look into the Objectives of the Housing Reconstruction Programme

From the outset of strategising the Bam reconstruction it was represented to have long term insights for tackling unsafe construction and the lack of attention to urban architectural quality and identity. However, organising purposeful programmes starts with the early important activity of setting objectives for projects and programmes (Chartered Management Institute, CMI, n.d.). Some unforeseen problems in the Bam urban housing reconstruction programme can be linked with its inadequate attention to the objective setting stage, for example, the lack of prioritisation and the absence of mutual understanding among the controller organisations. Multi-objective systems carry a potential conflict between high-level objectives (Hamdi 2010), thus, multi-objective systems require prioritising objectives (CMI, n.d.). If those priorities are not determined at the system formation stage, they will inevitably be determined during the programme implementation once a conflict arises.

The aforementioned bottleneck in design phase in urban housing reconstruction programme can be explained from this perspective: participant organisations in the programme were left on their own to interpret the priorities, which are naturally biased towards their disciplines. In principle, as presumed at the system formation stage, safeguarding cultural identity and architectural fabric and constructing earthquake resistant buildings were compatible because structural and architectural technicalities always cooperate in construction industry. Also, the objective on beneficiaries' participation was similar to people's role in a normal housing development procedure. But in practice, the priority for the KEO at the design phase was a conservative structural design as it perceived Bam as a special case. Contrarily, the priority for the Mother Consultancy was the traditional architectural features for which the optional additional seismic codes must have remained optional. Here the third objective of beneficiaries' participation added another crucial layer and pressing dimension of time to this dispute: bringing sudden demand, creating growing backlog and frustrated public, who already suffered from the disaster.

Arguably, the initial approach towards the objective on safeguarding architectural fabric was broad and idealistic. Aimed to make the most of the assumed 'window of opportunity' for change, the approach downplayed the post-disaster situation and the existing realities of housing development at the time. According to CMI (ibid.), objectives must be specific, achievable and realistic. They may be challenging but should not demand so much that the chance of success gets small (e.g. Adair 2008; CMI, n.d.). The initial architectural guidance and codes associated with this objective were directly extracted from traditional architecture in Bam and covered a wide range of architectural and urban elements: from the location of the new houses (to protect the garden city characteristics), to detailed architectural elements (such as raising the ceiling height of the hall) and forms, to extendibility of the house and climatic architectural features, room dimensions and proportions and so on. Such breadth was, indeed, a reflection of professionals' desire to inject qualitative criteria into quantitative everyday housing development practices which are unable to address architectural identity. The Bam case became a pilot for exercising a change and inform the comprehensive quantitative planning system at national level, since the idea was to establish councils similar to the BAUC for other cities too (interviews, 2013). Preliminary architectural workshops, organised for consultancy companies and locals, with focus on architectural perspective instead of linking those architectural expectations to costs, time and the framework of a housing reconstruction programme in a post-disaster and traumatic situation. Moreover, traditional materials could not be used in reconstruction because at the time seismic regulations for vernacular construction techniques were not developed. Too many architectural guidance on this objective without linking them to other organisational elements and the above realities led to too much expectations and in turn reduced chances of full delivery of this objective.

Undoubtedly, there is a necessity for embracing qualitative architectural features and addressing traditional identity in housing development—and in general the urban development planning system—and revisiting traditional housing typologies in practical terms. However, this objective tapped on Iran's fundamental problem of its established planning system. A housing reconstruction programme undertaken on individual owner-driven basis with individual contractor teams by a limited reconstruction fund within a complex post-disaster situation, is not the right place for a 'pilot project', although improvements are achievable. Logically, such improvements must target a few certain and most important issues, such as protecting the palm tree gardens and the garden city characteristics of the Bam area in order to simplify the organisational aspects of achieving the objective.

Such idealism is also linked to the BAUC's nature: a single-disciplinary architectural council with ambitions for a fast track change. Being a single-disciplinary council, it could not benefit from cross-disciplinary discussions and examinations which enhance deliverability.

17.8 Conclusion

The urban housing reconstruction programme in the Bam case can be considered as an innovative attempt for urbanising previous reconstruction policies previously and successfully implemented in rural areas in a historic urban context. It was through a systematic approach, based on the modification of urban housing process in normal situations and combining it with reconstruction policies and practical considerations to facilitate achieving the reconstruction objectives.

This final chapter of the book can be summarised in two directions: the first is about urban reconstruction and the second is a quest for change in urban development activities in normal situations. On the first direction, urban reconstruction, there are both highlights and shortfalls. Both highlights and shortfalls of the Bam's case for organising urban reconstruction carry valuable points for organising urban reconstruction in Iran and potentially international cases. Such attempt is one step forward for systematic thinking about large-scale urban reconstruction. For Iran, as a highly earthquake-prone country, this was the first of its kind but might not be the last.

The highlights include taking advantages of creativity, systematic approach and purposefully organising urban reconstruction. The essence of vertical cascading of strategic thinking helps to manage complexities of reconstruction for identifying required tasks, participant organisations, their relation and the workflow process with clear responsibilities and expected outcomes towards the objectives. The systematic approach links reconstruction policies and city regulatory procedures for housing, creating a modified housing development process throughout the city. Through exploring possible realistic improvements to existing housing development procedures for the reconstruction purposes and actively engaging and delegating accountability with local responsible institutions the reconstruction programme potentially becomes an agent for exercising change.

The programme's shortfalls, however, had direct implications on the measures for approaching the objective on safeguarding urban architectural identity during the programme implementation and contributed to unexpected conflicts and problems. Vertical cascading is necessary but not enough for organising a housing reconstruction programme strategically as it must identify and harmonise horizontal interconnections and interdependencies within the workflow process. Objectives should be prioritised and a mutual understanding about priorities before the programme implementation is vital.

Linked to other chapters in this book, the second direction was the quest for change and improvements to the status quo both in urban development practice and housing process in normal situations. Introducing three objectives in urban reconstruction in general and specifically in housing reconstruction programme was an urge for a change in urban planning system in normal situations: an urge to achieve safe construction and an appreciation of the traditional urban planning system. It was a demand for tackling existing failures in urban development activities. The way this programme was organised showed existing spaces for manoeuvring and improvement within the normal housing process. While a single post-disaster housing reconstruction programme is not a perfect pilot project for testing ideas on qualitative approach, there is no reason why further ideas should not be explored in normal situations. Promoting qualitative approach towards respecting Iran's infamous historic urban identity and integrating qualitative measures into existing housing processes might be difficult but certainly possible. Clearly exploring the possibilities of such improvements and changing the existing urban process and integrating necessary measures for them should not take another urban disaster in a historic city to happen. Both possible summarising directions, i.e. urban reconstruction and an urge for change in urban development activities in normal situations, called for action in contemporary practice of housing process and urban development. Time for action is right now.

References

Adair J (2008) The best of John Adair on leadership and management. Thorogood Publishing, London

Akhoondi A, Bahraini H (2000) Reconstruction management of the disaster-affected areas; the experience of reconstruction of housing in the earthquake-affected areas of the Guilan and Zandjan provinces, Iran 1990–1992. The Institute of Publication and Print of the University of Tehran, Tehran

Alexander DE (2002) Principles of emergency planning and management, illustrated. Oxford University Press, England

Alizamani MH (2012) Contemporary approaches in post-disaster housing reconstruction (unpublished Power-point presentation)

Ansoff H (1990) Implanting strategic management, 2nd edn. Prentice Hall, New York

Arefian F (2015) Organisational design and management for post-disaster reconstruction programmes; the case of Bam. (Ph.D. thesis), The Bartlett Development Planning Unit, UCL, London, UK (in process for book publication)

Armstrong M (2009) Armstrong's handbook of management and leadership: a guide to managing for results, 2nd edn. Kogan Page, London

Barakat S (2003) Housing reconstruction after conflict and disaster. Network Paper

Beheshti SM (1383) Bam, the world heritage (historical and cultural specifications of the Bam City in urban planning). In: Proceedings of the Bam we all want; technical and professional consultative workshop for reconstruction of Bam, Tehran, 25–26 Feb 2004

Benson C, Twigg J (2007) Tools for mainstreaming disaster risk reduction: guidance notes for development organisations

BHRC (2004) Initial urgent report of the Bam earthquake 26 Dec 2003. BHRC, Tehran

Bonabeau E (2007) Understanding and managing complexity risk. MIT Sloan Manag Rev 48 (4):62

Chartered Management Institute (CMI) (n.d.) Setting SMART objectives checklist 231

Comfort LK, Kapucu N (2006) Inter-organizational coordination in extreme events: the world trade center attacks, 11 Sept 2001. Nat Hazards 39(2):309–327. doi:10.1007/s11069-006-0030-x

Davidson C (2009) Multi-actor arrangements and project management. In: Lizarralde G, Johnson C, Davidson C (eds) Rebuilding after disasters: from emergency to sustainability. Spon Press, London, New York, pp 88–109

Davis I (2007) Learning from disaster recovery; guidance for decision makers. International Recovery Platform (IRP). http://www.undp.org.cu/crmi/docs/irp-decmakersleslearned-td-2006-en.pdf. Accessed 17 Nov 2010

Esmaeili A (2004) Advocating community demands in the process of reconstruction and urban planning for Bam. In: Proceedings of the technical and professional workshop for reconstruction and urban planning for Bam city. Ministry of Housing and Urban Development, Tehran

Ghafory-Ashtiany M (1999) Rescue operation and reconstruction of recent earthquakes in Iran. Disaster Prevention and Manag 8(1):5–20

Government of Iran (2005) National report of the Islamic republic of Iran on disaster reduction, Tehran. http://www.unisdr.org/2005/mdgs-drr/nationalreports/Iran-report.pdf. Accessed 22 Oct 2014

Hamdi N (2010) The Placemaker's guide to building community. Earthscan Publications, London, Washington, DC

Hanachi P (2004) New policies in urban planning and existing references in reconstruction of Bam. In: Proceedings the Bam we all want; technical and professional consultative workshop for reconstruction of Bam, Tehran, 25–26 Feb 2004

Hannagan T (2002) Mastering strategic management. Palgrave, Basingstoke Hampshire [u.a.]

Havaie MH, Hosseini M (n.d.) Bam earthquake: from emergency response to reconstruction. JSEE special issue on Bam earthquake

HFIR (2004) Housing reconstruction programme for damaged regions in the Bam earthquake

HFIR (2005) Guidance for reconstruction, residential, retail and public units of Bam, Baravat and surrounding villages, 2nd edn. Housing Foundation of Islamic Revolution, Bam

HFIR (2012) Documenting the reconstruction of Bam, Role and Function of Governments, International Agencies and NGOs in Reconstruction (focusing on Housing reconstruction)

Johnson C (2007) Strategic planning for post-disaster temporary housing. Disasters 31:435–458. doi:10.1111/j.1467-7717.2007.01018.x

Joodi M (2010) A brief review on Bam reconstruction program

Maskrey A (1994) Disaster mitigation as a crisis of paradigms: reconstructing after the Alto Mayo earthquake, Per. In: Varley A (ed) Disasters, development and environment. Wiley, Chichester, New York, pp 99–124

Meskinazarian A (2011) Social resilience of post-earthquake Bam. Department of Geography, King's College London, London

Mintzberg H, Ghoshal S, Lampel J, Quinn JB (2003) The strategy process: concepts, contexts, cases, 4th edn. Pearson Education, New York

Moghadam AS, Eskandari A (2004) Post-earthquake quick inspection of damaged buildings in Bam earthquake of 26 Dec 2003. In: Special issue on Bam earthquake, vol 5(4) (Winter 2004) —vol 6(1) (Spring 2004). International Institute for Earthquake Engineering and Seismology (IIEES), 2004

Morgan G (2006) Images of organization, updated edition (first edition 1997). Sage Publications, Thousand Oaks

Özerdem A, Jacoby T (2006) Disaster management and civil society: earthquake relief in Japan, Turkey and India. I.B. Tauris, London, New York

Pelling M (2003) Chapter 2: cities as sites of disaster. In: Pelling M (ed) The vulnerability of cities: natural disasters and social resilience. Earthscan Publications, London, pp 19–45

Quzai U (2010) Pakistan: implementing people-centred reconstruction in urban and rural areas. In: Lyons M, Schilderman T, Boano C (eds) Building back better: delivering people-centred housing reconstruction at scale. Practical Action Pub, Warwickshire, UK, pp 113–134

Saemian S, Erfanian Daneshvar A (2011) The process of technical control and supervision in reconstruction of Bam

Sartipi M (2006) Natural disasters and diagnoses of man-made factors, the cases of Manjil and Bam. Environ Sci 12:59–66

Shilderman T (2010) Putting people at the centre of reconstruction. In: Lyons M, Schilderman T, Boano C (eds) Building back better: delivering people-centred housing reconstruction at scale. Practical Action Pub, Warwickshire, UK, pp 7–33

Steering Committee for Reconstruction of Bam (2004) Decision by the President's special representatives

UN (2004) United Nations launches appeal for Iran earthquake URL. Available from http://www.un.org/News/Press/docs/2004/iha852.doc.htm. Accessed 10 June 2014

Wisner B (2004) Chapter 9: towards a safer environment. At risk: natural hazards, people's vulnerability and disasters. Routledge, London, pp 321–376

Zargar A (2004) Afterthoughts from an Iranian perspective. In: Mumtaz B, Noschis K (eds) Development of Kabul: reconstruction and planning issues. Comportements, Lausanne, p 1

Index

F.F. Arefian and S.H.I. Moeini (eds.), *Urban Change in Iran*,
The Urban Book Series, DOI 10.1007/978-3-319-26115-7

Z

Printed in the United States
By Bookmasters